START PREPPING!

START PREPPING!

GET PREPARED—FOR LIFE

A 10-Step Path to Emergency Preparedness so You Can Survive Any Disaster

By

Tim Young

selfsufficientman.com

COPYRIGHT

Disclaimer

Every effort has been made to ensure that the information contained in this book is complete and accurate. However, this book is sold with the understanding that the author is not engaged in rendering medical, legal, accounting, technical or other professional services or advice. Any ideas, procedures and suggestions contained in this book are not intended as a substitute for consulting with professionals. If you require such expertise, you should seek the services of a competent professional. The author will not be liable or responsible for any loss or damage allegedly arising from any information or suggestion in this book.

This book is for entertainment purposes only. The views expressed are those of the author alone, and should not be taken as expert instruction or commands. The reader is responsible for his or her own actions. Adherence to all applicable laws and regulations is the sole responsibility of the purchaser or reader.

For Liz and Maisy—The reasons I prepare

TABLE OF CONTENTS

YOUR FREE GIFT!

To thank you for your purchase,
I'm offering a FREE gift!

PLAYFUL PREPAREDNESS: Prepare your children—for life! is an eBook with 26 games for teaching preparedness, situational awareness and the survival mindset skills to children. It's a one-of-a-kind tool that can help your children become safe and self-reliant, whether you are there to protect them or not.

AND IT'S YOURS, FREE!
Download at selfsufficientman.com/playful-preparedness/

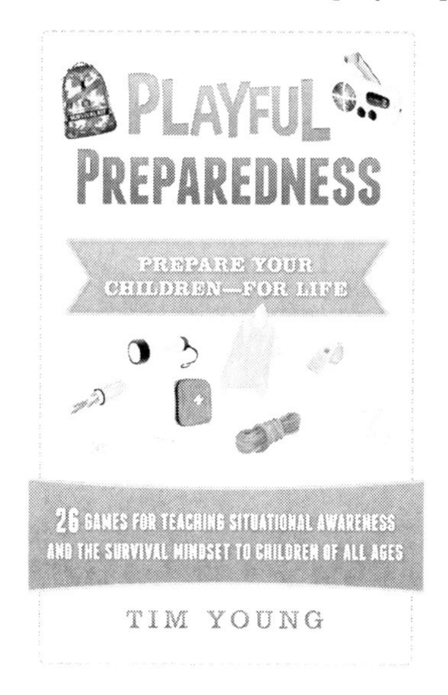

INTRODUCTION

On a chilly Sunday in December 2013, James Glanton packed his Jeep before setting off to frolic in the snow with his girlfriend and three young children. But while driving through Nevada's Seven Trough mountain range, he hit a slick patch that sent the group rolling down an icy embankment. No one was injured, but the Jeep landed upside down at the bottom of a remote, frigid ravine.

With no mobile phone reception and in bone-chilling temperatures, Glanton's survival mindset kicked in immediately.

Within minutes of the crash, to keep the family from freezing to death, he built a fire and kept it burning. As an avid hunter, he had packed a lighter, hacksaw and a magnesium fire starter in the Jeep. These resources allowed him to start a fire even with damp wood.

When the family failed to return home that evening, 200 volunteers began a search of the vast snow-covered mountains. With no success and temperatures plummeting to 16°F below zero, authorities suspended the search at 4:30 a.m. to refuel vehicles and recruit fresh volunteers. For two days, searchers combed 6,000 square miles of jagged, icy terrain in bitter cold, without finding a trace of the family.

Stranded, Glanton found his resourcefulness challenged. He began heating large stones in a campfire during the day and placing them inside the vehicle to keep the family warm at night. There was little to eat, but he did have a small supply of food and water in the vehicle. He carefully rationed it out, but still faced a crucial decision—whether to stay with his family or attempt to walk out and seek help.

His survival mindset would be put to the ultimate test.

As an outdoorsman, Glanton surely knew the life or death nature of the decision he faced—a choice that confronts many who become hopelessly stranded in the wilderness.

In a highly publicized tragedy just seven years earlier, television personality James Kim became lost in a fierce blizzard. After being stranded for seven days on an unpaved Oregon logging road, Kim, wearing only tennis shoes, a jacket, and light clothing, made his own fateful decision. He left his wife and daughters, a four-year-old and a seven-month-old, in his car while he sought help. He promised his wife he would return that day if he didn't find anyone.

Kim never returned.

Two days later a helicopter pilot spotted and rescued Kim's wife and two daughters as they attempted to walk out. Authorities continued the search for Kim, at one point finding clothing he had discarded, likely because he believed he was too hot. Paradoxical undressing is one of the final stages of hypothermia.

Despite a Herculean search effort that included repositioning a space satellite, 35-year-old Kim was found two days later floating in a stream of icy water. He was dead from hypothermia.

Five hundred miles to the east, it is not clear whether Glanton knew of the Kim tragedy. What is clear is that he made the decision to keep the family together at the vehicle rather than leaving them to seek help. He continued to seek solutions, even burning the Jeep's spare tire to create a white smoke plume in an effort to signal searchers. The family watched several times on Monday as planes flew high overhead, but searchers never spotted them. The family was forced to huddle in the Jeep for another frigid night as temperatures dipped to 10°F below zero.

On Tuesday morning, a volunteer using high-powered binoculars spotted wreckage. He and two others moved on foot toward the scene as helicopters were called in. The three searchers followed a trail of children's footprints in the snow until they came across tire tracks that snaked into a remote canyon. They followed the trail and found the Jeep, flipped upside down.

They also found Glanton, his girlfriend, Christina McIntee, and three children, all under the age of 10.

All were alive and huddled around a fire.

Doctors and authorities were shocked that the two adults and four children managed to escape the ordeal relatively unscathed, without even suffering frostbite.

> "The boys saw it as camping in the Jeep," Glanton said when he was interviewed on The Today Show.

Rescuers quickly praised the family's resourcefulness.

> "They stayed together and that was the key that allowed them to live through this experience," said Paul Burke, search-and-rescue coordinator for the state of Nevada. "You don't see that often in search and rescue. Staying together—that was a big deal."

Authorities proclaimed the survival of the Glanton family miraculous, but was it? Was it a miracle, or was Glanton *prepared* to make the most of his skills, his environment and his supplies to see his family safely through the ordeal?

The story had a happy ending, but I don't believe the outcome was the result of a miracle. Rather, Glanton made smart decisions, carried critical survival supplies, had the skills to do what was needed and displayed a keen survival mindset.

Glanton and his family survived because he was *prepared* mentally and physically for whatever came his way.

That is what prepping is all about.

A CULTURE OF UNPREPAREDNESS

In recent years, reality shows and news stories have thrust the word "prepping" into mainstream American consciousness. Creators of these shows knew they would draw a bigger audience by associating the term "prepping" with fearful catastrophic events such as widespread pandemic, global financial collapse, and doomsday prophecies.

To keep viewers and advertisers engaged, producers needed a fearless human icon to lead us through the pending doom. Someone viewers could both relate to and argue about around the water cooler so they would have an emotional investment in the show.

From the ashes of what prior generations simply called a survivalist, producers created the modern "prepper" to serve as the ultimate fearmonger.

Far from being cast as fearless superheroes, most of these made-for-TV preppers were selected based on a narrow set of criteria. The more paranoid, extreme and isolated a person was, the more eager producers were to feature them.

As the cameras rolled, these preppers strapped on guns, gas masks, and bug-out bags while constantly ranting that something was imminent—electromagnetic pulse, dollar collapse, polar shifts, super-volcanoes—potential catastrophes few had heard of before or thought about.

Audiences flocked to prime-time shows that featured camouflaged preppers burying food in ground caches and hiding in underground bunkers that were stacked with enough guns and ammo to start a small war. Some preppers even showed wide-eyed viewers how to filter and drink their own pee.

Viewers reacted predictably.

Preppers were ridiculed and labeled paranoid extremists. It was an understandable but unfortunate reaction, for the shows falsely portrayed the true essence of a prepared person's mindset.

I watched the shows, too, but accepted them for what they really were: entertainment.

Like most people who take preparedness seriously, I give no thought to zombies, prophecies, or asteroids. I don't own a bunker and—I like to think—will never drink my own urine, filtered or otherwise.

I do, however, give a lot of thought to ensuring I have what my family needs to survive. Namely, safety and enough food, water, shelter, money and skills so that we can thrive if times get tough, but especially if they don't.

I enjoy living in the moment, but I also think about what the future could hold and take steps to prepare. To me, that's just being a responsible husband and father.

But that kind of thinking means I get labeled as a prepper.

Evidently I'm not alone, for not everyone dismisses prepping as a crazy idea. Millions of Americans have now embraced prepping, and the number is growing with every natural disaster and act of senseless violence.

But the barrage of so many doomsday predictions paralyzes the majority of people. Rather than taking steps to prepare, they simply ignore the threats, essentially burying their heads in the sand.

As it turns out, sticking their heads in the sand is, at least metaphorically, a completely normal human reaction. After all, we're creatures of habit who enjoy predictable and secure lives. We believe that life will simply continue as it has.

This belief that life will continue as we know it is called "normalcy bias." You can also think of it as having your head in the sand, or the "ostrich effect".

Normalcy bias is simply the belief that tomorrow will be pretty much the same as it is today, and it has a firm grip on our psyche. When presented with sudden change, unprepared people seize up and the normalcy bias renders them unable to cope. As you'll see, at no time is this truer than when their lives are at stake, for normalcy bias gets people killed.

I believe it is also the reason why so many people fail to prepare for disasters and life-changing events.

Whether you have heard the term "normalcy bias" or not, we're all familiar with many tragic examples of this behavior.

Normalcy bias helps us understand why so many Jews continued living in Germany after the Nazis passed discriminatory laws against them and required they wear identifying yellow badges.

Even after the concentration camps were fully operational, few people were able to accept that a government had *actually* built gas chambers and ovens.

They had.

We see the same display of normalcy bias when large populations are asked to evacuate.

For weeks before the volcano erupted in 1980, park rangers issued warnings advising people to evacuate the Mount St. Helens area. Some residents ignored the warnings while sightseers and campers actually circumvented the barricades and entered the park. After all, they had camped there before and never experienced a disaster. This was their normalcy bias in action, which prevented them from assessing the very real potential danger. On the morning of May 18, 1980, the volcano erupted violently and killed 57 people.

While the term normalcy bias may be new to you, it has afflicted mankind for millennia.

Two thousand years ago, thousands of Pompeii citizens were buried in a thick carpet of volcanic ash from Mount Vesuvius. Instead of evacuating, they stood dumbfounded for hours watching the volcano erupt. Over 30,000 people died.

The same paradox affected survivors of the World Trade Center attack in 2001 when, according to a study from the National Institute of Standards and Technology, they waited an *average* of six minutes to *begin* leaving the doomed towers.

Elia Zedeño was one of those survivors. At her desk on the 73rd floor of Tower 1, she heard a booming explosion and felt the tower lurch as if it might topple. But her instinct wasn't to flee.

> "What I really wanted was for someone to scream back, 'Everything is O.K.! Don't worry. It's in your head,'" she said.

So she froze.

Fortunately, one of her colleagues reacted differently and screamed, "Get out of the building!"

Zedeño followed the command and made it out alive, but not before taking a moment to snag a mystery novel from her desk.

Others delayed much longer, up to 30 minutes, as they calmly turned off computers, collected possessions and, it seems, looked for others to tell them that everything would be okay.

They were held firmly in the deadly clutch of denial.

The problem is that "most people go their entire lives without a disaster," according to Michael Lindell, a professor at the Hazard Reduction & Recovery Center at Texas A&M University. "So, the most reasonable reaction when something bad happens is to say, this can't possibly be happening to me."

When faced with a horrible, jolting change, it seems our minds reject the bitter truth in favor of a palatable lie.

On December 7, 1941, a radar operator on Oahu reported seeing an unusually large "blip" on his radar screen. When he notified superiors, the reply he received was, "Don't worry about it," as authorities *assumed* it was a returning flight of U.S. planes. The large blip, of course, was the first wave of Japanese fighters and bombers, whose surprise attack on Pearl Harbor catapulted the United States into World War II.

Facing an imminent disaster, thousands of New Orleans residents who were *able* to evacuate refused to do so in advance of

Hurricane Katrina. After all, many had survived numerous storms amid false predictions that the levees could one day fail. So they stayed, and over 1,400 people paid for that decision with their lives. The unprepared survivors became homeless refugees, solely dependent on volunteers and authorities for food, shelter, water and medicine rather than their own preparations.

The paralyzing grip of normalcy bias hinders our ability to make decisions in all disasters, whether they are natural, manmade, or tragic accidents.

In 1977, a KLM flight slammed into a Pan Am plane in the Canary Islands, slicing it open. Passengers aboard the KLM plane died instantly, but most of the people on the Pan Am flight survived the initial impact. Once the plane came to rest, fleeing survivors ran past people who simply sat completely bewildered, but uninjured. Rather than exiting the plane they remained seated and watched with apparent disbelief for 60 seconds as flames encroached. They were unable to overcome their normalcy bias, which reassured them that "it" wasn't happening. "It" was, so 326 of the 396 passengers on the Pan Am plane also died, making the crash the deadliest accident in aviation history.

Most people believe that the chances of surviving a plane crash are slim, so it may surprise you to learn that the survivor rate in plane crashes is well over 90 percent.

The National Transportation Safety Board examined all air crashes that occurred between 1983 and 2000. Of the 53,487 people involved in those incidents, 51,207, or 96 percent, survived. They looked closely at 26 of the most serious accidents that involved fire and substantial damage. Excluding those in which no one had a chance, the survival rate in even the most "serious" accidents was 77 percent. So, even in *bad* crashes, three out of four passengers survive.

Though the facts indicate a high survival rate, people generally don't believe it. Ben Sherwood, author of *The Survivor's Club*, refers to this as the myth of hopelessness:

> "One dangerous consequence of the myth of hopelessness is that when people believe there's nothing they can do to save themselves, they put themselves in even greater peril. Before flying, they pop a few drinks in the bar. As soon as they get on the plane, they take off their shoes, crack open a book, read the paper, or crank up the iPod. They ignore the safety briefings and information cards. If the plane crashes, they figure it doesn't matter if they're drunk, barefoot, and blindfolded: They're dead anyway."

Since it's normal for planes to arrive safely, passengers simply don't prepare for anything other than what is normal. People generally take the same approach in all phases of their lives, which, I believe, is why we fail to prepare for emergencies.

Aviation safety expert Ed Galea studied over 2,000 survivor reports from aviation crashes. His findings clearly illuminate how normalcy bias costs lives when passengers try to escape a downed airplane.

> "People try and press a button on the seatbelt because in an emergency situation, they revert to normal behavior. And what is normal behavior for most people? Well, they experience a seatbelt in their car and in their car, it's a push-button system."

The European Transport Safety Council estimates that 40 percent of the fatalities in global plane crashes were *actually survivable.* So why did those victims perish instead of survive? In survivable crashes, experts say, it boils down to human factors and what you do—or don't do—to save yourself.

Case in point—The Pan Am crash in the Canary Islands.

The Pan Am passengers who sat bewildered were killed not by the air crash, but by their normalcy bias. Their minds couldn't accept the brutal reality that had been thrust upon them—that they were alive, but had to unbuckle their seat belts and exit the aircraft to stay that way.

Studies show that about 70 percent of people in a disaster are hindered by normalcy bias. Later in this book you'll read a story of a horrific fire that claimed 100 lives. I'll share a link to a raw video of that tragedy, which shows people calmly moving while surrounded by a blazing fire that, seconds later, claimed their lives.

Denial, delusion, and rationalization are *very* strong human biases that are difficult to overcome without a determined effort. Overcoming your normalcy bias requires a survival mindset.

OVERCOMING NORMALCY BIAS

By now you know that you have a 96 percent chance of surviving a plane crash, though I'm willing to bet that your normalcy bias is having a hard time accepting that.

Likewise, we're led to believe that if we suffer a gunshot wound we're done for. In truth, there's a 95 percent chance we'll survive if we get to a physician. Struck by lightning? Ninety percent survival rate. Venomous snakebite? Ninety-nine point nine percent survival rate.

To be sure, there's plenty to be fearful of in the world, but the first step to overcoming your normalcy bias is to deal with the truth and not the lies that we absorb through news, movies and propaganda.

Of course, it can be difficult to determine what the truth is.

From the 1950s through the 1980s, many Americans believed there would be a nuclear war between the Soviet Union and the United States. Some even built backyard bunkers to protect themselves from radiation fallout.

The cold war ended without conflict.

Today, many people are worried about a global financial collapse. It remains to be seen if that will occur and what the impact will be. Others obsess over plague-like diseases, maniacal terrorists, drones, and the National Security Agency reading their personal emails.

So as the year 2000 loomed, was it surprising that many people *really* believed computers would crash and their personal data—from financial to job records—would be lost?

To be safe, some took the plunge into prepping and stocked up on emergency supplies before the clock struck twelve on December 31, 1999. When computers continued working just fine after the Y2K hangover, these overnight preppers felt silly and went back to their normal way of life.

Given that many alarming predictions don't come true, it's understandable why so few people take the time to prepare for anything. Yet while mainstream media used Y2K and the 2012 Mayan prophecy to showcase (and ridicule) preppers, there's no denying this: hurricanes Rita, Ike, Sandy and Katrina *did* occur and kill thousands of people.

9/11 also really happened, as did the Fukushima nuclear disaster, devastating floods, raging wildfires, the 2007 financial crisis, tornado outbreaks, deadly superstorms, acts of terrorism, tragic school shootings, prolonged power failures and other disasters that tore families apart, destroyed homes, bankrupted businesses, killed thousands of people and cost trillions of dollars.

We are surrounded by so many deadly life-altering tragedies that ironically, we seem unaware of them.

These newsworthy events seem less real when absorbed through television and news reports—more like something from a movie than real, human suffering. Soon, the horrific images fade, normalcy bias takes over and we get on with our own lives.

Left unchecked, normalcy bias projects our current conditions into the future. Because we have never faced peril ourselves, we assume we never will. Therefore, we do nothing to prepare. When a disaster does occur—well, by now you know what happens. Odds are that we can't cope—and we may perish.

The fact is that small and large-scale disasters happen all the time, every day. And the truth is that they can happen to you. In fact, they probably *will* happen to you, sooner or later. You'll understand why when you read chapter two.

For the most part, you cannot stop these disasters from happening. You can only be prepared so that, when they do occur, you and your loved ones will come through unscathed.

You can attempt to contemplate every possible threat if you're so inclined, but it isn't necessary. Nor do you need to fully understand them all. It's okay if you don't want to take the time to understand the complex intricacies and probabilities of global financial collapse or coronal mass ejections (CMEs). You just need to understand that your life CAN change instantly.

For the worse.

Your normalcy bias has been telling you that these events won't happen, but are you willing to bet your life on it? Your family's life?

You can choose to be mentally and physically prepared for threats, or, like the ostrich, keep your head in the sand and pretend everything will be just swell. But I want you to be prepared.

My primary aim in this book is to help you break free of your normalcy bias so that you and your family will not only survive, but thrive in uncertain times. It may not be a widespread pandemic—it may simply be a house fire where action can save you but mental paralysis will kill you.

You can't control what life throws your way. You can, however, control how you prepare and react.

Overcoming normalcy bias requires making conscious choices and not relying solely on authorities, who are often poorly equipped to anticipate or respond to disasters anyway. In other words, overcoming normalcy bias requires mastering a survival mindset.

This book will help you to do just that.

HOW TO USE THIS BOOK

I wrote this book because I'm worried about you and the millions of people who are ill prepared to face a crisis.

I suspect that some of you, if not most of you, are rather new to prepping. Perhaps something happened to you or you became aware of a concern that opened your eyes to the risk your family faces by not being better prepared.

Often, when a person realizes how woefully unprepared her family is for any kind of disruption to their modern way of life, she's engulfed by quiet panic.

Some families react like caged tigers. If they have the resources, they start buying their way out of their lack of preparedness by stocking up on freeze-dried foods that they've never tried, supplies that they don't know how to use and, in extreme cases, purchasing bunkers to bury underground.

This is not a book about doomsday prepping. It is a book about practical preparedness. So let's just take it slow, one step at a time. Prepping is not about panicking—the time to panic, if there is one, is when disaster strikes and you're not prepared.

So let's get you prepared!

While some of the topics we'll explore are alarming, I won't attempt to frighten you or use scare tactics. My goal is simple; to free you from your normalcy bias and motivate you to take your first steps to becoming prepared, or, if you've already begun, to go much further.

And I'm not just talking about being prepared for "typical" prepper events, such as natural disasters. I'm talking about being prepared for everyday life, such as random violence and job losses.

I plan to guide you not by suggesting that a worldwide disaster is imminent or even likely, but rather by helping you to see the real risks that we all face, and walking you through how to prepare for them. Chapter two describes nearly 30 potential crises that we are likely to face, from death of a spouse to financial crisis and, yes, even pandemics.

This book is both an easy read and, if you're new to prepping, an exhaustive resource. As I hope you can see from this introduction, you won't find this to be a dry textbook full of lists and conspiracy theories. Rather, you will learn the many commonsense reasons why it's important to take personal responsibility for preparedness and not rely solely on others.

I wrote this book to help anyone become more prepared, but I am particularly aiming to motivate those who have no preps at all, or are relatively new at preparedness. Perhaps you've resisted because, you tell yourself, you're too busy or don't have enough money or space. If you're like most people you may simply think it won't happen here—it won't happen to you.

If you're already a master prepper, you'll find some great advice and inspiration within these pages, and you'll have the perfect gift to give to your non-prepping friends and family; this book!

Checkpoints & Challenges

To help you get the most out of this book, I've included Checkpoints in each chapter. These are designed to allow you to pause for a moment, reflect on what you just absorbed and think about your own situation. I felt this was important because there is a lot of information presented in this book that may be new to you. Just as we need time to digest a big meal, I wanted to give you opportunities to digest what you're learning and, most important, to think about how it applies to your own situation.

As an example, here's your first Checkpoint:

CHECKPOINT—RATE YOUR NORMALCY BIAS

- Numerous scientists and risk assessors such as Lloyd's of London consider a severe geomagnetic storm to be almost inevitable in the future.
- Such an event may create an electronic magnetic pulse that could destroy our power grid for over a year.
- How strongly do you agree or disagree with these experts?

In addition to Checkpoints, I included several "Prepper Challenges" at the end of the book in Appendix I. By taking the time to perform these challenges at your leisure you'll increase your skills and confidence at the same time.

THE 10-STEP PATH TO PREPAREDNESS

Prepping is about much more than food and water storage. There are many aspects of preparedness that are equally critical; some are actually far more important than food storage. As you'll see, few people give these areas any thought until it's too late.

To help you understand the importance and interrelationship between these preparedness areas I've created a visual aid and an outline. This tool, which I call the Ten-Step Path to Preparedness, will serve as the framework for this book.

This graphic illustrates that being prepared is much more than simply buying a few cases of beans and water. It's a series of interconnected actions that will enable you to go about life quite normally during a common disaster, and to survive a widespread catastrophic event.

In the following chapters you'll find many details within each of these 10 steps, but at a high level the steps to preparedness are:

Step 1: Water

Step 2: Shelter, Warmth & Comfort

Step 3: Food

Step 4: Energy

Step 5: Sanitation & Medical

Step 6: Personal Security & The Survival Mindset

Step 7: Money & Documents

Step 8: Communication

Step 9: Travel Security

Step 10: Skills

This book will teach you that preparedness is about much more than having *stuff*. It's about having *skills* and, most important, embracing the *survival mindset*.

While I hope you embrace preparedness and, over time, amass the skills and supplies necessary to survive a long-term crisis, I am focusing the content of this book to help you get started. I want you to be prepared, at a minimum, for a short-term disaster, *which I define as up to one month*. Therefore, I'm not addressing certain self-sufficiency topics that are near and dear to my heart, such as gardening and animal husbandry.

By the time you finish this book you'll learn what many others and I already know—that prepping is not irrational and pessimistic. Rather, prepping may be the most rational and optimistic gift you can offer your family.

With every path, there's a first step, and the path to getting prepared is no different. Take your first step now and remember—the day before most disasters is a calm, ordinary day.

Just like today.

Trust yourself, not the government. Too many people have for too long placed too much confidence and trust in government and not enough in themselves.

Ron Paul

CHAPTER ONE

LAND OF THE FREE, HOME OF THE UNPREPARED

I was born in the 1960s; not *that* long ago. Civilized Americans then enjoyed most of the same conveniences we take for granted now, such as electricity, stocked grocery shelves, indoor plumbing, telephones, and air travel.

My family was different.

Until I was four years old my family lived in what I now refer to as the "old house." It was a wooden, four-room cabin perched on a mountain stream that my grandfather built in the 1940s. Actually, grandpa not only built it, he also cut down the trees and used his sawmill to finish the lumber for it. When the freshly cut timber later dried, wide cracks opened throughout the cabin. The cracks permitted bedside stargazing in the summer, but grandma covered them with cardboard in winter to keep the chill at bay.

The old house bore no resemblance to the fancy cabins that now grace the cover of glossy magazines. Grandpa simply provided his family a rural homestead complete with an outhouse and pigpen, but without running water, electricity, or a telephone. Certainly without cable television.

When we wanted a drink we walked to our spring, filled a bucket and toted it home. If she wanted hot water, grandma had to start a fire first. Our house was pitifully small and our garden was remarkably large, measured in acres.

My sister and I played outside, in the dirt. Sometimes we made toy guns out of sticks. If we got out of line, we got "whooped" with a leather belt, so we made a point to not get out of line. Not once did Child Protective Services come to complain about the dirt, the toy guns, or the spankings.

When she wasn't filling her books with S&H Green Stamps, Grandma cleaned clothes on a washboard in the creek and wrung them dry. She churned butter and sewed on a treadle sewing machine, her feet creating a rhythmic beat to complement the dueling banjos, guitars, and mandolins my uncle and our neighbors played on the porch.

After harvesting the garden she shucked corn on the same porch and cooked dinner on a wood stove in the kitchen. It goes without saying that we all survived the kitchen heat and the oppressive north Georgia summers without fans or air conditioning.

All that changed by the time I was four, when my family built a brick home complete with all the modern conveniences of the day. Telephones replaced the need for folks to come "sit a spell," and there was no porch to sit on anyway. With electricity powering window fans, a front porch was unnecessary in the fancy new house.

Like most people, my parents and grandparents were lured by luxuries that promised an easier life. They were told all they had to do was stop *producing* what they needed to survive. Instead, they simply needed to earn a paycheck so they could *consume* what they needed to survive. Someone else would produce it for them.

Fast-forward several decades and we now find a movement of people who long for the perceived virtues of a simple life—the kind of life our family had while living off the land.

Grandma, like so many others at the time, seemed only too happy to leave that life behind. And with that transition from producers to consumers, our society sentenced future generations to a life of dependency.

I understand why we "upgraded," but it saddens me. Today the "old house" exists only in a few torn and faded Polaroids.

Nowadays we're all so busy with our lives that we're only vaguely aware of how our species rose to the top of the food chain. We humans mastered remarkable survival skills, such as how to hunt, grow, and preserve food, clothe ourselves, make and control fire, build shelters, repair our bodies and invent useful tools. We handed down and refined those skills through countless generations.

And then, we stopped.

Now we're going backwards and are losing those skills FAR faster than we acquired them.

From Baby Boomers to Millennials, we have rejected the survival skills handed down to our ancestors that they were supposed to pass on to us.

Most people today have virtually no natural survival skills, are completely unfamiliar with (and allergic to) the natural world that surrounds them and are unable to protect their families and their possessions in times of crisis.

Ask pretty much anyone if they can make soap, cut down trees and build a house or snare a rabbit and you're likely to get a blank look. And then be asked to leave.

For instance, how many of us have the skills that our *recent* relatives, such as my grandparents, had? How many of us are used to doing the following?

- Saving seeds, year-round gardening, and foraging for wild foods.
- Owning surplus canning jars, a pressure canner and knowing how to preserve food without electricity.
- Having weapons and the skills to use them for hunting, trapping, and self-defense.
- Making and reloading ammunition.
- Harvesting timber and building shelters.
- Making shelters and fire from natural materials.
- Farming, harvesting, and handling livestock.
- Milking animals, making cheese and butter.
- Making and knowing how to use natural medicines.
- Butchering animals for meat.
- Saving and tanning hides.
- Sewing and mending clothes.
- Knowing how to make lye, soap, lotions, and salves.
- Making, sharpening and repairing hand tools.
- Possessing a keen survival mindset.

Those survival skills required virtually no money—only time and practice. Now ask yourself, can you do the things to *survive* that don't rely on the dollar or electricity?

How comfortable would you be without HVAC, running water, transportation, or the knowledge that a state-of-the-art medical facility was just a free ambulance ride away?

How would you feel if you woke up tomorrow and you had none of that? No dollars, no electricity, and no assistance? Your normalcy bias tells you it can't happen, but since history shows that it can, how would you feel if it did?

Really think about that for a moment.

IT HAS HAPPENED BEFORE

History reveals that unexpected disasters happen with regularity. Some are easily understood, such as natural disasters. Others are often perplexing, such as government collapse, hyperinflation, and severe space weather.

Unfortunately, as German philosopher Georg Wilhelm Friedrich Hegel said 200 years ago:

> "We learn from history that we do not learn from history."

In our hectic, modern world, we're so rushed to and fro that we can hardly remember what day it is, much less what we ate for dinner the night before last. It goes without saying then that most people have short memories, and this is particularly true when it comes to understanding and learning from history.

If you are a student of history then you know that financial collapses and bank holidays have occurred. For those who didn't pay attention to the subtle clues, when they did happen, they seemed to happen overnight. Amid the devastation that leaves the overwhelming majority of the population with less wealth, fewer opportunities, economic hardship and anxiety, it becomes clear that the clues were there all along.

Ninety-nine percent of you reading this book did not live through the Great Depression 80 years ago. Still, we all know it happened because of the profound effect and hardship it had on so many people around the world. We're aware of that effect some 80 years later because of the global hardship that ensued, and with today's interconnected global markets, the potential for widespread suffering is far greater, and the suffering will occur at lightning speed.

WHY WE DON'T PREPARE

Many people don't know where to start when it comes to prepping—so they never start at all. They may know what prepping means (to be prepared), but are unsure *what* they should prepare for. After all, recommendations often call for a hand-crank radio, a full tank of gas and enough food and water for three days. But what if you're preparing for a widespread pandemic where gathering in public would be ill advised? Stockpiling for three days would be woefully inadequate.

It's ironic that relatively few Americans embrace preparedness when studies show that we are, in fact, fearful of potential disasters.

A 2014 Chapman University survey asked 1,500 Americans to rank their concern of natural disasters. Topping the list of fears was tornadoes and hurricanes, followed by earthquakes, floods, pandemics, and power outages. But when asked if they were prepared for a disaster the *overwhelming* majority of Americans responded "NO."

A 2013 poll by FEMA found similar results.

> "Americans are mostly unprepared for natural disasters, even those in the most vulnerable areas."

It turns out that there are a LOT of people living in vulnerable areas.

A staggering 91 percent of Americans live in places at a moderate-to-high risk of earthquakes, volcanoes, tornadoes, wildfires, hurricanes, flooding, high-wind damage or terrorism, according to the Hazards and Vulnerability Research Institute at the University of South Carolina.

I find these data fascinating, for Americans have a long tradition of being resilient and fiercely independent, but now it seems a culture of unpreparedness and dependency smothers the land of the free.

With our lives quite literally on the line, why do so few take personal preparedness seriously?

The answer is that our normalcy bias assures us that "it" won't happen to us, or that there's nothing we can do anyway.

I call this "preparedness paralysis."

In many instances, people believe they will simply dial 9-1-1 in an emergency. That's unfortunate, for even with working phone service, it turns out that most people's expectation is way off as it relates to response time. They expect it will be an hour, but in reality, response times range from three days to never during a widespread crisis.

These people live in a state of denial, and will suffer the consequences should disaster befall them.

> "There are four stages of denial," says Eric Holdeman, director of emergency management for Seattle's King County. "One is, it won't happen. Two is, if it does happen, it won't happen to me. Three—if it does happen to me, it won't be that bad. And four—if it happens to me and it's bad, there's nothing I can do to stop it anyway."

Who knows for sure why more people don't take prepping seriously, but I know the reasons I often hear.

1. "Hey, if it's my time, it's my time."
2. "If anything happens law enforcement and emergency responders will handle it."
3. "I can't afford to prep. I can barely make it as it is."
4. "I live in a small apartment. There's no space for extra supplies."
5. "I'm too old to start."
6. "I have faith in God."
7. "I have faith in the government."
8. "I don't know where to start or what to prep for. I've heard so many scary scenarios, I'm confused."
9. "I see it on TV but it won't happen to me."
10. "I'm too busy...I don't have the time."
11. "No one I know preps."

The reasons people give for not preparing are as puzzling to me as the prepper's mindset is to non-preppers, who dismiss preppers as fringe lunatics.

Fringe lunatics or not, here's a certainty we can all agree on: There *will* absolutely be more natural disasters, and some will be catastrophic.

Even if you agree with that statement—and what sane person wouldn't?—you run the risk of not taking action to prepare unless you break free from your preparedness paralysis.

Consider this excerpt from a 2006 Time Magazine article titled *Floods, Tornadoes, Hurricanes, Wildfires, Earthquakes ... Why We Don't Prepare.*

> "A serious hurricane is due to strike New York City. Experts predict that such a storm would swamp lower Manhattan, Brooklyn and Jersey City, N.J., force the evacuation of more than 3 million people and cost twice as much as Katrina. New York City ranks second on a list of the worst places for a hurricane to strike, but in a survey measuring the readiness of insured home-owners living in hurricane zones, New Yorkers came in second to last."

This article, written fresh on the heels of Hurricane Katrina, appeared six years before Superstorm Sandy pummeled New York City and the northeast. It presciently warned of the potential dev-astation awaiting New Yorkers and highlighted the need to pre-pare.

As is usually the case, few gave it serious thought.

While not a hurricane when it made landfall, Superstorm Sandy claimed 159 lives and caused $65 billion in damage in the U.S., making it the second-costliest weather disaster in American his-tory, behind only Hurricane Katrina.

CHECKPOINT—WHY YOU DON'T PREPARE

- Examine the reasons you have resisted preparing or haven't done more.
- Rewrite your logic.
- **Example**: I don't prepare because it's expensive.
- **Rewrite**: I will lose a lot of money if a disaster happens and I'm not prepared to save my food and valuables.

WHAT DOES IT MEAN TO BE PREPARED?

You may recall the saga involving James Glanton from the opening of this book. His family was put to the test not by a doomsday event, but by their own actions and the forces of nature.

Their survival story dispels a common misconception relating to prepping. The misconception is that prepping is mainly about preparing for cataclysmic doomsday events. It's not—it's about being prepared to survive *whatever* disaster befalls you. And, as you'll see in the next chapter, the disasters you are most likely to face are not widespread, but rather isolated to you and those around you.

I carry a "Survival Bag" in my vehicle along with an emergency car kit at all times. A medical kit, tent, food, water, fire starters, tools and other provisions are in each of our vehicles, just as a car jack is along with the knowledge of how to use it.

You'll learn in this book why I take those precautions, but in a nutshell, that simple act is *what it means to be prepared.*

Preparedness isn't just about having survival *stuff*—it means considering in advance the potential dangers you may face and preparing yourself to not be a tragic victim when an unexpected crisis visits you.

It means to *not* assume that the ATMs will work, that the grocery shelves will always be stocked and that the electricity will always flow.

It means taking the time to contemplate, in advance, how you would handle certain situations. "What if my car breaks down? Should I listen closely to the flight attendant instructions—what if the plane crashes? What if a fire engulfs this restaurant—do I know where all the exits are? What if my region loses power for weeks—how would I get food and water?"

Perhaps most important, being prepared means having a survival *mindset* and situational awareness of your surroundings.

Prepping isn't for weirdos or the super rich—it's a personal responsibility that we all have—to our families and ourselves. I believe it's also a social responsibility we share. After all, the more people who are prepared, the less impact a disaster has.

The good news is that you're likely already a prepper, although you probably don't know it.

YOU MIGHT BE A PREPPER

Despite the dramatic and sensationalized portrayal by shows such as National Geographic's Doomsday Preppers, being a prepper is more sensible than you may think.

For example, and with inspiration from comedian Jeff Foxworthy, if you do any of the following—you just might be a prepper.

- Carry a spare tire in your car.
- Grow your own vegetables or have fruit trees.
- Have a life insurance policy.
- Have a savings account.
- Have ever practiced an earthquake, fire, tornado or evacuation drill.
- Keep a plunger by the toilet.
- Wear a seat belt while driving.
- Take a diaper bag when going out with the baby.
- Hunt or fish.
- Save money for retirement.
- Keep extra batteries on hand.
- Have fire extinguishers, flashlights, or emergency candles in your home.
- Go to the dentist for a check-up or get a physical from the doctor.

Why would you take such precautions? You purchase car insurance, but are you planning on being in an accident?

Why have money saved that you don't need today? Why save for retirement if you don't know for sure you'll be around to enjoy it?

Isn't it inconvenient and costly to get exams when you're perfectly healthy, or carry diaper bags, spare tires, back-up batteries and insurance policies that you don't need *right now*?

Sure it is, but you do it for good reason. You want to be prepared so you can *enjoy* life rather than be inconvenienced or face a crisis unprepared. *And that makes you a prepper.*

If you hunt or fish, you have a prepper skill.

If you garden or preserve food, you have a prepper skill.

If you can start a fire, you have a prepper skill. And that's good news because prepping is just as much about having *skills* as it is about having *stuff*.

Prepping is simply being able to take care of your family without relying on outside assistance, which likely won't be there when you need it most anyway.

So like it or not, virtually *all of us* are preppers to some degree, as were each of our ancestors. Unfortunately, most of us have simply forgotten how to put food by and prepare for the proverbial rainy day.

The question we each have to answer is, "where do I want to draw the line?" And, we each have to draw a line because none of us can be totally self-sufficient in the purest sense of the phrase. We must rely on others or things for some of what we need—unless you plan to perform your own open-heart surgery.

GET STARTED ON YOUR PREPAREDNESS

Even though I mentioned some possible disastrous events that could happen, the truth is that you're most likely to experience something more personal and devastating to you, but not to others. I'm thinking of family injury or disability, job loss, house fire, severe weather and so on. The following chapter details a thorough list of likely disasters that you should consider preparing for.

Remember that while *your* personal misfortune may not result in hardship for others, for you the impact could be the same as a global financial collapse. So it's prudent to prepare for whatever the future holds.

While there is a sense of inherent urgency to prepping (after all, we don't know *when* the disaster will happen, now do we?), I'd like to encourage you to catch your breath and take it one step at a time. You didn't get to your current *lack* of preparedness overnight and there's no way you can learn all the skills you need to be prepared by tomorrow.

Sure you can buy stuff, and I will encourage you to do so if you're able. To help you in that regard, I've provided hyperlinks for many products I use and recommend in the eBook version of this book. Click on the links ONLY if you're interested in seeing the product on Amazon.com since that is where I purchased most of mine. Otherwise just ignore the links and continue reading.

If you don't have access to the eBook version, you can find a full list of the products and resources I recommend online at self-sufficientman.com/recommendations.

Even if you purchase products for your preps, you still need a preparedness plan and you must hone your survival mindset. Mastering the survival mindset is one of the most important takeaways from this book, so I've devoted all of chapter eight to explaining how you can learn it and teach it to your children.

Now that you know what preparedness is really all about, proudly embrace prepping rather than distancing yourself from preppers. After all, the famous motto of the Boy Scouts is, "Be Prepared." No one questions that or dismisses Boy Scouts as paranoid, and for good reason. **For why would you not want to be**

prepared to take care of yourself and your family?

It makes sense when we tell the Boy Scouts to be prepared and it makes just as much sense when we heed the advice ourselves.

The more people we encourage to stock their pantries, learn preparedness skills, and become prepared to protect their families, the less social unrest and widespread panic will come our way when times get really tough.

As they always do and inevitably will.

Now that you have an understanding of WHY you should prepare, let's examine the disasters that may be headed your way.

AUTHOR'S NOTE

This book will be WORTHLESS to you if you simply file it on your bookshelf with the intention of only reading it *after* an emergency strikes.

This is NOT a survival manual to be used after the need.

This is a preparedness guide that will help you and your family if you follow its recommendations *before* the need.

CHAPTER TWO

LIKELY DISASTERS

We are creatures of habit. As a result, one of the hardest things for a person to do is to anticipate sudden change. Instead, we extrapolate our past to project our future.

Because modern society in developed nations is remarkably stable, this approach generally works well. But sometimes we are blindsided by traumatic events that unnerve us and threaten our way of life.

These include sudden financial market declines, pandemic scares, food shortages and coordinated terrorist attacks that create a wave of destruction throughout our political, military, financial, social, and food systems.

The idea that our way of life could suddenly change, or even disappear, is absurd to most people. We briefly look to history to confirm it hasn't happened recently and go about our business. We disregard pandemics of 100 years ago, plagues of the Middle Ages and countless empire collapses as tragedies that may have afflicted humanity *then*, but could not possibly *now*.

This time is different, we tell ourselves, as we dismiss dire predictions from society's fringe that disasters lay straight ahead.

But are we merely the captain of our own Titanic, steering our family's ship full-speed ahead into a disaster that could be averted?

For the disaster is often not the event itself, but the impact it has on humanity. An impact that could be greatly lessened, if not outright avoided, if people were simply prepared.

While doomsday preppers love to talk about any number of scenarios that will create The End Of The World As We Know It (what preppers call "TEOTWAWKI"), the truth is that each *single* scenario is statistically unlikely. This is why many people ridicule those doomsday predictions and continue about their business, failing to make any preparations at all.

Even though the occurrence of any single widespread disaster may be unlikely, in a group of dozens of possible calamities it becomes very likely that a society-changing disaster *will* occur.

When that happens most people will be blindsided, only to see, *if* they survive, the warning signs clearly visible in retrospect.

What *is* likely is that you will face some type of disaster that has the potential to be personally devastating to you and your family.

If you're new to prepping then don't start by preparing for an unlikely asteroid to wipe out civilization. Rather, start practically with events that are, I'm sorry to say, much more likely to impact you. While they have the potential to be traumatic when they do, you and your family will survive much more comfortably if you take the time to prepare NOW.

That's the point of this book and of being a prepper.

The list of hazards and tragedies one is likely to face will vary, of course, from location to location and from person to person. For instance, a person living at the base of a volcano has a different natural disaster priority than someone living on an earthquake-prone fault line or living in a major river flood zone.

To encourage your own thinking, here is a list of 27 personal and societal risks, in *my* order of most likely to least likely.

TYPES OF RISKS

Risk Number 1: Job Loss—If you and/or your spouse lose your job or jobs tomorrow, how will you pay for food, utilities, health care and other necessities to maintain a secure lifestyle?

Risk Number 2: Death/Disability—It's nearly impossible to prepare for the sudden loss of a close family member, or the debilitating disability suffered by a loved breadwinner. If there is any solace to be taken it may be that you're physically and fiscally prepared to bear the burden due to your preparedness.

Risk Number 3: Local Natural Disaster—The list of *local* natural disasters includes

- floods,
- wildfires,
- small or rural earthquakes,
- mudslides and landslides,
- winter storms,
- wind storms,
- severe heat waves,
- severe droughts that cause *locally* higher food costs, wells to dry up and threaten fire suppression,
- hurricanes and
- isolated tornadoes.

Risk Number 4: House Fire—There are nearly 400,000 home fires each year in the U.S. Do you have an escape plan? Can you trust that your children will follow it?

Even if you have smoke detectors, their life expectancy is about 10 years. Do you know how old yours are? Do you have a system for testing them and replacing batteries?

If you have preps put aside, would you lose all your hard-earned preps in a home fire?

Risk Number 5: Home Invasion/Crime—According to the latest FBI statistics there is a property crime every three seconds, a burglary every 15 seconds, a VIOLENT crime every 26 seconds, a forcible rape every six minutes and a murder every 35 minutes.

Are you prepared to defend yourself, your family and your possessions?

Risk Number 6: Family Crisis—Mom or dad needs long-term care or to move in with you. Are you prepared financially, physically and otherwise to handle an unexpected family need?

Risk Number 7: Widespread Natural Disaster—In addition to the local natural disasters, widespread disasters include

- strong earthquakes that hit high-density population centers or logistically critical zones,
- severe drought that drives nationwide food costs much higher,
- widespread tornado outbreaks,
- extreme winter storms such as Superstorm Sandy and
- powerful hurricanes.

Risk Number 8: Prolonged Utility Failures—When a prolonged power failure occurs, how will you heat your home? How will you cool your home? What will you do with the food in your refrigerator and freezer? How will you get potable water without electricity? How will you drive if fuel stations are closed? How will you get cash if ATMs are not working?

Risk Number 9: Neighborhood Dangers/Increasing Crime Rate—Look no further than Ferguson, MO in 2014 or Baltimore, MD in 2015 for "newsworthy" riots. In your area there may be escalating crime rates that aren't newsworthy, but pose a real danger to you and your family.

Risk Number 10: Shooting at a Public Place—When I was a kid it was unheard of to have a shooting at a theater or a school, but today it's not. Do you and your children have the situational awareness to be prepared in such an event?

Risk Number 11: Pandemic Outbreak—Deadly diseases have wiped out large populations repeatedly throughout history, and with today's travel-centric society, disease can spread far faster than authorities can react to or than you can prepare for. Pandemics can affect local, regional, national or global populations.

Risk Number 12: Widespread Civil Unrest, Rioting or Collapse of Social Order—Unfortunately, this often accompanies other crises such as food shortages, terrorist attacks, natural disasters, economic collapse and so on. Areas of high population density are particularly vulnerable.

Risk Number 13: Terrorism—Both domestic and international terrorism is increasing. The risk includes use of biological weapons, EMPs, chemical/HAZMAT incidents, attacks on the power grid and federal properties, cyber attacks, etc.

Risk Number 14: Hazardous Materials Incident—They happen. For example, in 2014 a chemical spill poisoned the Elk River, leaving 300,000 people without drinking water in West Virginia.

Risk Number 15: Bank Holiday or Economic Collapse—If you think that the risk of a severe economic crisis is an idea to be dismissed, consider what Jamie Dimon, Chairman and CEO of JPMorgan Chase, had to say in his April 2015 letter to shareholders.

> "**There will be another crisis**. Triggering events could be geopolitical (as in the 1973 Middle East crisis), a recession where the Fed rapidly increases interest rates (as in the 1980-1982 recession), a commodities price collapse (oil in the late 1980s), the commercial real estate crisis (in the early 1990s), the Asian crisis (in 1997) or so-called "bubbles" (the 2000 Internet bubble and the 2008 housing bubble)."

Regardless of the trigger, the interdependency of today's global markets means that events in Asia and Europe can create a domino effect that plunge the U.S. economy into recession or depression.

Alternatively, an economic crisis may not come in the form of a "crash" but rather a prolonged economic decay where everyone becomes desperate for money, including the government.

Risk Number 16: Cyber Terrorism—In 2015, Hackers linked to China stole over four million U.S. government workers' social security numbers and personal identifying information.

With a war on cash driving modern societies to 100 percent digital economies, the risk of personal financial loss is a very real threat.

Risk Number 17: Massive Power Grid Failure—Few of us acknowledge how fragile the power grid is that makes modern American life possible. A simple solar storm could disable it for years.

Beyond natural disasters that could take down the power grid, a headline from a 2014 Wall Street Journal article warned *The Growing Threat From an EMP Attack*. The subtitle stated coldly, "A nuclear device detonated above the U.S. could kill millions, and we've done almost nothing to prepare."

Risk Number 18: Widespread Food Shortage—Food shortages could develop quickly for any number of reasons, including power failures, drought, gas or water shortages, pathogen contamination, transportation strikes, earthquakes, hurricanes, economic woes, military conflicts, and disease.

Risk Number 19: Nuclear Disaster—This is of particular concern if you're close to a nuclear reactor or within fallout range based on prevailing wind patterns. Of course, winds don't always blow the way they're supposed to, so most of us are at risk to some degree.

Risk Number 20: New Madrid Earthquake—In 2014, the U.S. Geological Survey updated its seismic hazards map and listed the New Madrid fault line among the highest risk of earthquakes. The fault stretches through the southern and midwestern U.S. A strong earthquake in that region would cripple much of our transportation system, which is what fuels our modern food system among other things.

Risk Number 21: Dam Failure— This threat is far down on my list, but it's a serious concern if you live near a dam.

Risk Number 22: Volcanic Ash Fallout—Also far down on my list, but if you live near a volcano, watch out.

Risk Number 23: War and Terrorism—It seems we're constantly at war, somewhere. As horrific as armed confrontation is, other than soldiers, Americans don't seem to worry about their personal risk. This is likely because the battles seldom take place on our soil, unless you count terrorism as war—which it is. It's the worst kind of war, an invisible kind that can strike anywhere, anytime, and spares no one. Domestic youth, religious zealots, organized foreign terrorist groups or anyone in between can perpetrate it. Often we don't know until it's too late to act.

Risk Number 24: Martial Law—Martial Law means that the military has taken over a specific area of the country and that area is now governed under the law of the military. Whatever the cause, Martial Law means that you'll have reduced mobility and very limited access to normal supplies and services.

Risk Number 25: Super-Volcano—Every few million years or so, "super-volcanoes" emerge that can erupt continuously for thousands of years. When any of us encounter the phrase "every few million years" it tends to invite ridicule, which is why this risk is rated so low on the list. Yet, the risk is real, and the most alarming potential super-volcano to Americans is located beneath Yellowstone National Park.

Whereas the risk from a volcano eruption (Risk #22) is limited to an area around the volcano, the effect of a super-volcano eruption would be widespread and catastrophic. When it happens, the ash and volcanic gases from a super-volcano will *wipe out most living things* over a widespread area.

Risk Number 26: Abrupt Climate Change—The science fiction film *The Day After Tomorrow* made for great entertainment, but could the Earth really undergo a sudden climate shift? A July 2015 headline from the Washington Post suggests so, with European scientists claiming a "Mini ice age likely."

Oh well, the only prepping plan I have for this is to switch from SPF100 sunscreen to wool blankets. You can't prepare for everything.

Finally, Risk Number 27: Asteroids and Zombies—Okay, well—they deserve a spot on the list, but don't spend much time worrying about or preparing for them—other than watching Shaun of the Dead to learn how to use vinyl records as weapons against Zombies.

I believe this list of 27 threats is more than enough for you to contemplate as you start prepping. However, some preppers have a much longer list, fearing everything from polar shifts to Biblical Armageddon. My list is now your list, so feel free to edit as you see fit—as long as you think seriously about the risks you face and the consequences of being unprepared.

For me, I believe the potential occurrence for the first 18 threats is great enough to warrant my preparation. Actually, I've taken measurable steps to prepare for all other than number 27, which is where I draw the line. I'd rather enjoy the moment than prepare for an asteroid impact. As for zombies—well—I'll just pick up what tips I can by watching The Walking Dead.

Even though I arranged the list in *my* order of probability, that is, of course, subjective. After all, you may be wealthy or comfortably self-employed, so the risk of job loss may not be near the top for you.

Some events that I ranked lower in personal likelihood still demand your attention. For instance, I placed a U.S. nationwide bank holiday and/or economic collapse at number 15, but I *absolutely* recommend you prepare for it as best you can nevertheless.

You may not consider financial collapse to be likely, but then again it isn't statistically likely that you'll lose your house in a fire. But you still have homeowner's insurance, don't you? And that gives you peace of mind.

Similarly, you should make preparations against financial collapse and other potential disasters.

Sadly, these calamities can and often do strike in tragic combinations. As awful as it is to endure, it's not unfathomable to imagine a disaster destroying a house and killing a parent who was also the breadwinner, delivering three or more catastrophes at once.

There's no doubt that prepping requires a commitment of both time and money, but there's a huge payoff. Once you're prepared, you'll be free to enjoy life today with the comfort of knowing your family can survive tomorrow if the unexpected occurs.

DON'T PREPARE FOR SPECIFIC EVENTS

Now that I've shared with you my list of possible threats, I'd like to encourage you, as a beginning prepper, to *NOT* prepare for any specific event. Doing so is a common mistake that new preppers make.

After all, what good is it to have prepared for a tornado when a flood strikes instead, or to prepare for a pandemic but not for a (much more likely) loss of income?

With that warning out of the way, let's take our first step onto the 10-Step Path to Preparedness, so that you can start prepping now!

"Prepping is not living in fear of what might happen. It's living with confidence regardless of what happens!"

Tim Young

CHAPTER THREE

PREPAREDNESS STEP 1: WATER STORAGE & PURIFICATION

To many of us, the smell of licorice is sweet and conjures memories of a playful youth. But some chemicals such as MCMH produce a similar aroma. For many West Virginians, the smell recalls the 2014 chemical spill in the Elk River that poisoned the water supply.

Rebecca Roth, a Charleston, WV mother, was pregnant at the time.

> "My daughter was not yet two and I was pregnant with my son—I went from feeling happy for what the new year held to being filled with fear for our family's health. Being afraid that I wasn't keeping my children safe and healthy is one of the worst fears I have ever experienced."

Like 300,000 other West Virginia residents, Roth suddenly found herself without water to drink after 10,000 gallons of a toxic coal-washing chemical spilled upstream from a water treatment plant that serves the area surrounding the state capital of Charleston.

> "After the water crisis, my life centered around making arrangements for drinking, cooking, dishwashing, laundry and bathing without using tap water," Roth continued.

People rushed stores for bottled water as soon as authorities announced the spill. Within an hour frantic residents emptied the store shelves amid reported fistfights and scuffles. For days, authorities instructed residents to flush the chemicals from their home plumbing systems by continually running faucets.

> "We tried our best to keep our family safe, but it was hard with conflicting information." Roth said. "Afterwards we learned that children shouldn't have been around during the flush and that the windows should have been opened, information that we didn't have at the time. We had no idea how dangerous this stuff is that got into our water. What worried me most was how the chemical spill was affecting the fetus growing inside me."

It took 48 hours for authorities to begin dispersing water, which first went to hospitals and nursing homes. For up to 10 days, residents had had no running water to drink, take showers, wash clothes or wash dishes.

A year after the spill, Roth was relieved but worries about the future lingered.

> "When my son was born in August, I was relieved to count 10 fingers and 10 toes, but not all signs of health are so easy to see. It will be years before we know if the reproductive health of either of our kids has suffered at all because of the chemical spill."

Clean water, like clean air, is something that most of us grew up taking for granted. But if access to either is hindered, it can be deadly.

New preppers often think first of storing food. While having food reserves is critical, it is down the list of priorities in an emergency situation.

An axiom in the survival community is the Rule of Threes. This simply states that, in a *harsh* environment, a typical person cannot survive for more than

- three minutes without air,
- three hours without shelter,
- three days without water, and
- three weeks without food.

Of course, some people can exceed these survival levels, but if you haven't had a drink in three days or a bite to eat in three weeks, you'll be desperately seeking one—that is, if you're still alive.

Even though air is the first priority, I'll make the assumption that you'll have air to breathe, since this is a guide to getting prepared rather than a guide to living on the moon.

And even though shelter comes before water in *extreme* survival situations, I believe water comes first in virtually all disasters.

There are many ways to approach each step in the path to preparedness. The choice that's right for you will depend on what you are preparing for, where you live, your budget, available space and so on.

Given these variables, I will describe what you must try to achieve, but I will include a detailed list of options you can choose from. Ultimately, it's up to you to comprehend the big picture and make choices that are right for your situation.

As it relates to meeting your needs for potable water, there are three methods to consider: bulk storage, filtering, and disinfection.

Unless you have a lot of available space, it's unlikely that you'll be able to store all the water you need in a crisis that extends beyond a few weeks. Therefore, I recommend layering all three approaches by having *some* water stored in addition to having the ability to source, filter, and disinfect water.

WATER STORAGE

On average, humans are composed 60 percent of water. So a 200-pound person carries 120 pounds of water, every day. The fact that water is both heavy and bulky to store creates a challenge for preppers.

Modern life allows us to twist a faucet and enjoy a seemingly never-ending supply of life-sustaining water that comes from—somewhere. But any number of disasters can cut off that supply, from extended power outages to drought and chemical spills. When those disasters strike, you'll have plenty of things to worry about.

Let's not make access to potable water one of them.

Start by putting some water aside. The more the better, so let's examine all the different ways you can store water.

Storing water can be as simple as purchasing water jugs from your local grocer or filling containers that you already have. If you choose to package water yourself, be sure to use only food-grade containers.

Food-grade plastic buckets and drums designed for water storage work well, but I prefer using smaller containers such as two-liter bottles. They're ubiquitous, easy to store and convenient to toss into a vehicle should you need to evacuate. If you're worried about tampering at all, you can even purchase replacement screw caps to ensure they haven't been opened when you go to use them.

Another option that some preppers favor is Water Brick storage containers. These BPA-free stackable containers hold 3.5 gallons each for long-term water storage. They can also be used to store dry foods, such as rice, oats or beans, allowing them to serve multiple purposes.

When relying on used containers, be sure to clean, sanitize, and thoroughly rinse them prior to use. You can make a sanitizing solution by adding one teaspoon of liquid household chlorine bleach to one quart of water. Just remember to use only household bleach *without* thickeners, scents, or additives. Treat each gallon of non-chlorinated water with eight drops of liquid household chlorine bleach when filling the container.

Water from a chlorinated municipal water supply can be added without treatment to clean, food-grade containers.

Many experts suggest storing one gallon of water per adult, per day, but I suggest you aim for two gallons.

Using my recommendation, a two-week supply works out to 56 gallons for a family of two adults. The math is simple: 2 people X 14 days X 2 gallons/day. That's not a month's worth, but you can't get to a month without first getting to two weeks, so it's a good start.

I suggest two gallons because you'll likely need more water than you think. In addition to drinking water, you may need to reconstitute dried foods such as pastas, freeze-dried meals, rice and beans. You'll also need water for cleaning dishes and utensils, and personal hygiene. Of course, that's just for people—if you have pets and livestock you'll need to consider their needs as well.

YOU WANT ME TO STORE HOW MUCH WATER?

Before we continue, let's take a moment to discuss where to put all that water. After all, the 56 gallons is only for two adults, and is only for two weeks. You may have a family of two adults and two children with a goal of storing a month's worth of water. In that case you'll need space for at least 160 gallons.

Regardless of whether it's 56 or 160 gallons, it sounds like a lot of space, I know. Just catch your breath for a moment and I'll show you how you can easily achieve it.

One of the options I employ is to use a heavy-duty 5-shelf Sterilite unit. These are those plastic freestanding shelf systems you often see at Home Depot or Lowes.

We can fit 16 one-gallon jugs of water on each shelf, for a total of 80 gallons in one shelving system. It only takes a floor space of about 24 inches x 36 inches, something pretty much anyone can find room for in a laundry area, garage, or even a closet. This cost-effective solution will ensure you have 80 gallons of drinking water when you need it. Adding a second unit would meet the monthly needs of a family of four.

For very limited storage space, consider an inexpensive WaterBob emergency water reservoir liner. These liners fit into standard bathtubs and can be filled with up to 100 gallons of fresh drinking water. They come complete with a handy siphon pump for dispensing water as needed. At less than $30 as of this writing, a WaterBob doesn't require additional space (if you already have a bathtub) and offers a lot of peace of mind.

If you have a basement or garage, you may be interested in a water storage tank. One I can recommend is the Sure Water brand, which is available either as a 260 gallon or 525 gallon container. These tanks can also be stored just fine outside since the high-density polyethylene material is UV stable and used for outdoor water tanks.

If you do store one outside just be sure that the water doesn't freeze, for if it does you'll have to wait until it thaws before you can access it. They'll be far too heavy to move easily.

There's a lot to like about the Sure Water tanks—except the price, which, as of this writing, is $600 and up. But if you have the budget and space for it, you could store enough water to get you through most emergencies you're likely to face.

Whatever option you choose, be sure to store water only where potential leakage would not damage your home or apartment.

If you find yourself in a pinch, remember that you may already have lots of water stored without knowing it. A standard hot-water heater contains 40-50 gallons of water. To remove the water, turn off the electricity or gas that powers the heater. Locate the draining valve at the bottom of the tank, which should have a spout to which you can attach a potable drinking hose. Once you have removed the water, filter and disinfect as described in the next section.

Another potential source of water can be found in waterbeds. Waterbeds hold up to 400 gallons, but some contain toxic chemicals that are not fully removed by many filters. If you designate a waterbed as an emergency resource, be sure to drain it yearly, refill it with potable water and add two ounces of bleach per 120 gallons. This way you'll have an excellent potential supply of water without taking up any additional space in your home. If you're at all uncomfortable with or unsure about the quality of the water, always disinfect it before using it.

If you have one, residential in-ground swimming pools can have over 20,000 gallons of water. But preparedness experts are torn on whether you can safely use pool water for drinking.

The reason it's debatable is because the content and quality of pool water varies greatly from one house to the next. Some pools are salt water, while others rely on chlorine.

Another issue is how well the pool is maintained. When disaster strikes, is the water pristine with a fresh injection of chlorine, or is it cloudy with algae?

These factors combine to render it impossible to make a sweeping statement that swimming pool water can or cannot be used safely.

FEMA states that it is unsafe to use swimming pool water for emergency purposes, adding that, "chemicals used to kill germs are too concentrated for safe drinking."

However, while some owners use fungicides and chemicals, others do not. So can those people safely drink the water?

In the following section on water filtration, I describe the Big Berkey water filter system, which *will* remove the chlorine from swimming pool water. While the manufacturer claims that the filter *will* also extract harmful chemicals such as herbicides and pesticides, they acknowledge that some types of harsh chemicals in the pool *may not* be filtered out.

Then again, you're likely already drinking chemicals. In a 2013 study, German researchers analyzed water samples from 18 commercial producers and found more than 24,500 different chemicals lurking in a *single* bottle of water.

The bottom line is *you* need to know what chemicals are in your pool. If you're unsure, follow the FEMA recommendations and use the water as a source for personal hygiene, cleaning and related uses, but not for drinking.

Now that I've covered the importance of storing water and the ways you can go about it, don't become obsessed with longer-term water needs just yet. Remember, this book is about how to *get prepared*, so for now—**just put some water aside**.

CHECKPOINT—WATER STORAGE

- Figure out how much water your family requires for two weeks. Remember to include any extended family members you plan on caring for.
- Calculate how much space you have to store water. Don't forget all nooks and crannies, such as under beds, closet floors and shelves, rain collection barrels, etc.
- Figure out what is reasonable for you to store and what storage options are most sensible for you.
- Write a summary of your current state, what you would like it to become and how you will get there.
- **Example:** I need 84 gallons of water stored for my family for two weeks. I will fill 40 recycled gallon jugs and keep them on a shelf in my garage. I also have 50 gallons stored in a hot water heater for a backup source, and I know how to access and purify it.

Longer term storage of water is challenging due to its weight and the space it requires. Few people find it practical to store more than a couple of weeks' worth of water, which is why the ability to filter and purify water is so critical.

WATER FILTRATION

After your stored water runs out in an emergency, you likely won't be able to find any to purchase. Therefore, you need the ability to collect and filter more water, or the ability to disinfect it, which is covered in the next section.

The first step is to identify sources of water from which you can draw, and the best time to do that is before a disaster occurs. Simply make a note on a physical map of streams and lakes that you can access. If you have the tools and know-how that I'll describe in a moment, you can access water from lakes, streams, ponds, rain barrels, and even muddy puddles.

The good news is that water is virtually all around those of us who don't live in the desert. The bad news is that it must be treated. Otherwise, you and your family face infection.

Water filtration ranges from removing visible debris to filtering out harmful bacteria.

In my opinion, a must-have, portable product is a **LifeStraw** Personal Water Filter. At only about $20 each, these lightweight drinking straws can be kept in vehicles, survival bags and around the house.

While they›re not useful for transporting bulk water, they are lifesavers when it comes to making water potable from pretty much any source. Just lean over a pond or stream and use the LifeStraw to sip directly from any water source, as it removes over 99.9 percent of waterborne bacteria and protozoan parasites.

We keep five-gallon buckets on hand to transport larger volumes of water. Water is heavy, and carrying more than this can result in injuries at a time when medical help may be unavailable.

You should keep at least one clean food-grade bucket available for this purpose.

Gravity Filtration/Purification

Once the water is collected, the next step is to filter it. While it can be filtered with a cloth and then boiled to disinfect it, that requires heat. I prefer to have a high-quality gravity filter that renders water safe to drink without using fuel.

A product I rely on to ensure my family has access to clean drinking water is the Big Berkey water filter and purifier. These filters remove nearly all contaminants with a 99.99 percent reduction in viruses and have a flow rate of up to 192 gallons per day. The pores within the water filter elements are so small that pathogenic bacteria are simply unable to pass through.

The Big Berkey is a very powerful filtering system that can purify untreated raw water from lakes, streams, puddles and stagnant ponds. There are many published lab studies of the effectiveness of Big Berkey filters in removing a high percentage of fluoride, nitrates, arsenic, chlorine, pharmaceuticals, coliforms, BPA, pesticides, heavy metals and more from water. You can see links to the test results at tinyurl.com/watertests.

Unfortunately Big Berkey filters are not inexpensive and cost up to $300. Regardless, I consider them essential as they can provide thousands of gallons of safe drinking water, even if the only water source is a muddy puddle.

Before using the Big Berkey or similar water filter, you should pre-filter the water to remove as much suspended silt and solids as possible. You can use a cotton shirt, clean coffee filters, cheesecloth, or a bandana for that purpose. It is important that the cloth used is clean, as dirty cloth may introduce additional pollutants. If you use this cloth repeatedly, be sure the same side is facing up each time so as not to introduce sediment into the water. Dirty water or water with large suspended particles will clog the filter more quickly, so using this pre-filter approach will extend the life of the water filter.

There are other ways to filter water and you can build your own if you have the time or lack the money to purchase a Big Berkey. Just do a web search for "DIY Water Filters" and you'll find many tutorials and videos.

Without a filter/purifier such as the Big Berkey, you need the ability to *disinfect* the water at *all* times, *wherever* you are. The good news is that this is pretty straightforward.

WATER DISINFECTION

Filtration systems such as the Big Berkey can dramatically reduce the contents of harmful bacteria but none of them can guarantee the *complete* removal of germs. Disinfection is the process that ensures drinking water is free from harmful organisms or pathogens.

There are several common purification methods for water that can be used prior to consuming water. These measures will kill microbes but *will not* remove other contaminants such as heavy metals, salts, most chemicals, and radioactive fallout. Distillation is required to remove salts, heavy metals, and chemicals.

NOTE: There is a key difference between **biologically** contaminated water and **toxic** water.

Biologically contaminated water contains microorganisms such as Giardia, bacteria, or viruses that can lead to infections.

Toxic water contains chemical contamination from pesticide runoffs and so on.

Boiling, filtering, or chemically treating water can remove or kill microorganisms from biologically contaminated water, *but will not remove chemicals* from toxic water.

Therefore neither filtering nor disinfecting would have *removed* toxins in the water from the Elk River spill—that would have required distillation. This underscores the importance of having stored potable water.

Disinfecting Water With Purification Tablets

Many campers are familiar with Potable Aqua iodine-based water treatment tablets. Easy to use and carry in a vehicle, purse and backpack, these tablets, which are intended for short-term emergency use only, render most water bacteriologically suitable for drinking.

The chief advantage to using the Potable Aqua tablets is that they are inexpensive, come in a small package, will produce potable water in 30 minutes and require no heat or effort other than collecting the water.

A disadvantage is that they are only effective for small volumes of water and are not considered as safe as boiling or treating water with chlorine.

Also, be aware that some people are allergic to iodine and cannot use it as a form of water purification, and some people allergic to shellfish are also allergic to iodine. Pregnant women are advised against using iodine. For these people, I recommend a filtering system such as the Big Berkey, boiling, or chlorine disinfection.

Disinfecting Water by Boiling

Boiling is the most certain way to kill all microorganisms, but it requires heat, which may be a challenge during certain disasters.

Prior to boiling, cloudy or dirty water must first be strained through a clean cloth to remove any sediment or floating matter. After you allow the filtered water to settle, draw water from the top where it will be cleanest. Then, boil the water.

How long you boil the water depends on whom you listen to.

While the World Health Organization (WHO) suggests boiling the water for 20 minutes, the Wilderness Medical Society says that water temperatures above 160°F will kill all pathogens within 30 minutes, and temperatures above 185°F within a few minutes. So in the time it takes for the water to reach the boiling point all pathogens will be killed, even at high altitude.

The CDC agrees that boiling for only one minute is fine, stating, "Water should be brought to a rolling boil for 1 minute. At altitudes greater than 6,562 feet, you should boil water for 3 minutes."

Aside from the high-energy costs to boil, another disadvantage is that boiling alters the taste of water since air is released from water in the process. Aerating the water by vigorously stirring or pouring back in forth between containers re-introduces oxygen and improves the taste.

Solar Disinfecting of Water

According to the WHO, ultra-violet rays from the sun can inactivate and destroy pathogens in water. Simply fill transparent plastic containers with water and expose them to full sunlight for five hours, or for two consecutive days under a 100 percent cloudy sky. The outdoor temperature is irrelevant.

The plastic bottles should be no larger than two liters and must be clean. This is a good use for soda bottles if you're a soda drinker; just clean the bottles after use and store with your water supplies.

Solar disinfection is achieved by a combination of radiation and thermal treatment. If water temperature of least 122°F is achieved, an exposure period of one hour is sufficient. Solar disinfection requires *clear water to be effective*, so be sure to pre-filter and draw clear water from the top before pouring into plastic bottles.

A great way to ensure the water bottles are heated is to lay them flat on a reflective surface, such as Mylar bags or Mylar blankets, aluminum foil, or sheet metal.

Disinfecting Water With Chlorine

Chlorine is a chemical widely used for disinfecting drinking water due to its ease of use, availability, and relatively low cost. If you're concerned about adding chlorine to your drinking water, know that it is common for municipal water sources to add chlorine or hypochlorite to drinking water. It kills all viruses, certain bacteria and is used to stop the spread of waterborne diseases, such as cholera, dysentery, and typhoid.

To be effective, chlorine must be added in quantities sufficient to destroy all the germs but not so much as to affect the taste adversely. The chlorine must also have sufficient contact time with the pathogens—at least 30 minutes. Be sure to use *only* bleach without thickeners, scents, or additives.

Before disinfecting with chlorine pre-filter water if it is cloudy to remove sediment. Use a clean dropper (add to your emergency preps if you don't have one) and add bleach to water in the following ratios:

- 2 drops bleach per quart of water.
- 8 drops bleach per gallon of water.
- .5 teaspoons bleach per five gallons of water.
- 5.5 teaspoons bleach per 55 gallons of water.

If the water is cloudy, colored, or *very cold*, double the amount of bleach.

After adding bleach, stir the water well and wait 30 minutes. The water should have a slight chlorine odor. If it doesn't, repeat the dosage and let stand for another 15 minutes before use.

If the chlorine taste is too strong, pour the water from one clean container to another and let it stand before drinking.

If the water remains cloudy, repeat until the water is clear to drink.

Bleach Shelf Life

While chlorine bleach is effective at purifying water, it degrades over time. Consider the following statement from a Clorox Bleach representative.

> "We recommend storing bleach at room temperatures. It can be stored for about six months at temperatures between 50°F and 70°F. After this time, bleach will begin to degrade at a rate of 20 percent each year until totally degraded to salt and water. Storing at temperatures much higher than 70°F could cause the bleach to lose its effectiveness and degrade more rapidly."

Since bleach becomes less effective over time you'll need a plan to rotate bleach in and out of your inventory, or you'll need an alternative to commercial bleach.

CALCIUM HYPOCHLORITE—AN ALTERNATIVE TO BLEACH

An alternative to using commercial bleach is to create your own bleach as you need it. High-Test Calcium Hypochlorite (HTH) is marketed for water treatment and is commonly sold as pool shock.

It is available in a granular form at any location that sells pool supplies, has a 10-year shelf life when stored in a cool, dark place and removes a variety of disease-causing organisms such as bacteria, yeast, fungus, spores, and viruses.

When purchasing HTH, look for 65-70 percent **pure calcium hypochlorite with nothing else added**, such as anti-fungals or clarifiers.

How to disinfect water using calcium hypochlorite

For your personal protection, work in a ventilated area and wear eye protection. If HTH comes in contact with certain substances it can spontaneously combust—refer to the next section on calcium hypochlorite safety.

Follow these steps from the Environmental Protection Agency website for Emergency Disinfection of Drinking Water with HTH:

- Add and dissolve one heaping teaspoon of granular HTH (approximately ¼ ounce) for each two gallons of water, or 5 milliliters (approximately 7 grams) per 7.5 liters of water.
- This will produce a stock chlorine solution of approximately 500 milligrams per liter. **LABEL THIS "DO NOT DRINK."** THIS STOCK SOLUTION IS YOUR BLEACH.
- To disinfect water, add the **stock chlorine solution** in the ratio of one part of chlorine solution to each 100 parts of water to be treated. This is roughly equal to adding 1 pint (16 ounces) of stock chlorine to each 12.5 gallons of water or (approximately ½ liter to 50 liters of water) to be disinfected.
- To remove any objectionable chlorine odor, aerate the disinfected water by pouring it back and forth from one clean container to another.

You can view the EPA's guidelines for using calcium hypochlorite at tinyurl.com/mysafewater.

While the EPA instructions do not mention this, I recommend using HTH when the water temperature is 75°F or warmer since chlorine's effectiveness at killing pathogens diminishes rapidly at lower temperatures. In general, chlorine and all sanitizers work best at temperatures between 75° and 120°F. If you need to increase the temperature, put the water in direct sunlight.

A little bit of HTH will go a very long way, as a one-pound bag can disinfect over 20,000 gallons of drinking water. Therefore, a one-pound bag is all you need in your preps, though you may wish to store more for barter and charity.

Finally, be sure to have a printed copy of how to purify water using any of the methods described above. If you're reading this as an eBook or listening to an audio version of this book, *write them down*!

CALCIUM HYPOCHLORITE SAFETY

As a white powder, calcium hypochlorite certainly looks harmless enough, similar to standard laundry powder. **However, it can be deadly if handled improperly.**

According to the CDC, calcium hypochlorite decomposes rapidly on contact with acids and produces chlorine and oxygen, resulting in a fire and explosion hazard. It is especially volatile with petroleum by-products such as brake fluid, oil, ammonia, turpentine and other substances commonly stored in garages and workshops. All it takes is a leak or an accidental spill to create a horrible tragedy.

In 2002, a passenger vehicle with a box of calcium hypochlorite in the rear next to an engine-cleaning product violently erupted into flames. Somehow these products combined as the family of five rode along an Oregon highway. When the products mixed they *ignited instantly* into a horrific fire that engulfed the entire vehicle. While the parents were just able to exit the vehicle, they were not able to rescue all their children, even though they received severe burns trying. One child was saved, but two children died in the fire.

That isn't the only vehicular tragedy involving calcium hypochlorite.

In 2000, a Kentucky woman suffered severe burns in a freak explosion inside her car after purchasing calcium hypochlorite and an algaecide. The chemicals leaked from their containers and combined, creating a flammable substance that ignited in her car. The injured woman spent eight months in a hospital after the intense flash fire, ultimately losing her eyelids, lips, and ears.

While both of these tragedies are horrific, I recommend that you learn from them rather than being fearful of storing valuable prepping supplies.

Many supplies that preppers store, such as ammunition, fire-starting supplies, lye, fuel...even pressure canners and more *can* be dangerous, *but are not* when handled properly.

Here is what the CDC has to say about safely handling calcium hypochlorite.

- Follow product label directions for chemical storage.
- Keep pool chemicals separate from all acids and other household items.
- Keep chemicals dry and don't mix different chemicals (for example, different types of chlorine products).
- Lock chemicals away to protect people and animals.
- Keep chemicals cool in a well-ventilated area away from direct sunlight.
- Keep chemicals closed in original, labeled container.
- Don't leave pool chemicals in a car for long periods of time, especially on hot days.

- Store in an area without drain or sewer access.
- Keep separated from food and feedstuffs.

Always remember—the best source of drinking water during an emergency is clean water you have stored with your emergency supplies.

CHECKPOINT—WATER FILTRATION & DISINFECTING

- Make a list of water sources in your area. If a bug-out location is part of your emergency plan, be sure to list water sources in that location and en route to the retreat.
- Mark the water sources on a physical map.
- Decide how you will transport the water from each location.
- Decide how you will pre-filter water (T-Shirts, cheese-cloth, paper towels, etc.)
- Determine which method of disinfection you prefer and if anyone in your party is allergic to iodine.
- Write a summary of your current state, what you would like it to become and how you will get there.
- **Example**: I will purchase two 50-gallon rain barrels and set them up to catch rain water from my roof. I will purchase calcium hypochlorite, store it safely and print written instructions on how to use it. I will practice disinfecting water with calcium hypochlorite before the end of this month, and I will keep one LifeStraw in each vehicle and survival bag.

WATER TREATMENT FLOWCHART

I've presented a number of water treatment options that may be new to you and I know that can be confusing. To help you visualize the options available to you, I've created a flowchart to illustrate the choices and processes for rendering water safe to drink.

The flowchart below lists some water sources that authorities such as FEMA *do not recommend using* for drinking water, such as swimming pools, waterbeds, and even water heaters. **I am not recommending that you should use them either.** However, I'm presenting them because, in a disaster, your water options may be limited unless you have a lot of stored drinking water. This chapter and flowchart are designed to help you make the *best* choice you can if your options are limited.

FINAL THOUGHTS ON WATER

Remember, it takes electricity to pipe water to faucets, so no electricity means no running water. Regardless of the methods you use, aim for a mix of water storage solutions. If you have the space, much of your emergency water can be stored in large containers, such as in the Sure Water tanks, but be sure to keep enough water stored in portable containers. The need for this will become apparent when we cover survival bags and evacuations later in this book.

Always have at least one backup method for water purification in case one fails. This can be any combination of methods, such as a water filter and storing HTH to disinfect water.

Even if electricity and water does flow, contamination of your water may occur at the source after natural or man-made disasters. This is not uncommon in the event of hurricanes, earthquakes, floods and tsunamis.

To be prepared to meet your family's needs for water, take the following steps:

- Store as much water as possible, and continually monitor the supplies for leakage.
- Stock more water than you anticipate needing.
- Maintain the ability to purify water at home and while traveling.
- Know where you can source additional water supplies (water heater, lakes, streams, etc.).

Be careful to manage consumption of salty foods during extreme water shortages since salty foods will increase your thirst.

This is a fine line to walk, however, as sodium is a required element that must be replaced, since your body loses it through sweat and urine.

SaltStick Caps could be a valuable item for you to stock. The electrolytes in the capsules minimize heat stress and muscle cramping due to perspiration. They are ideal for preppers who may have limited access to water, as well as anyone who lives in a hot place and tends to dehydrate.

Finally, don't forget your pets. Have a source of water for them along with bowls regardless of whether you're at home or on the road.

Nature doesn't care. And for that reason, we must care.

Kathleen Tierney, Natural Hazards Center at the University of Colorado

CHAPTER FOUR

PREPAREDNESS STEP 2: SHELTER, WARMTH & COMFORT

There are several reasons why it's important for you to have adequate shelter during a crisis. For one, shelter can help to keep your family safe. Many emergency situations involve a heightened risk if you interact with others, such as civil unrest, looting, or pandemics. Having shelter creates a physical barrier between you and other people, who may be dangerous or ill.

A solid shelter also helps to deter wildlife. Depending on where you live this may include common invertebrates (disease-carrying mosquitoes and cockroaches, scorpions or venomous spiders), mammals (bears and dogs) or reptiles (venomous snakes).

Of course, shelter is critical to protect your family from the elements. After all, staying dry is the first rule of survival and maintaining body temperature in a comfortable range is crucial.

By having a shelter that protects you from other people, animals, and the elements, you will not only have time to rest and recuperate, you will have a more positive state of mind and be able to concentrate on the challenge confronting you.

There's a two-part dilemma, however.

First, what happens if you're not at your permanent location when a disaster hits. What do you do then?

Second, what happens if a disaster causes you to bug out and leave your permanent shelter?

To address these challenges, I'll walk you through two types of shelters: permanent and non-permanent. I'll then drill down within non-permanent to examine multiple alternatives.

We'll also look at making the crucial decision of when to stay and when to bug out.

PERMANENT SHELTERS

Permanent shelter structures include houses and fixed-location buildings. Hopefully, you have a house or apartment so you can shelter-in-place during a disaster. For many reasons I will discuss, you may not want to count on that, which is why access to a designated bug-out location is so important.

Bunkers or in-ground storm shelters are also options, but—let's be honest—few of us have access to these luxuries, so these may not belong as part of your preparedness plan.

If you do have a roof over your head, then you may be able to stay put during many types of widespread emergencies. For example, I can't imagine a hurricane, blizzard, pandemic, or economic collapse requiring me to leave my rural property, given my location.

Localized emergencies such as isolated tornadoes, house fire, mudslides, or wildfires and the like are horrific, but since they affect only you or your local community, you will likely be able to find support from your state or region.

In my case, we're surrounded by thousands of acres that are mainly planted forest. If a wildfire burns out of control, I can absolutely imagine having to leave—and fast. So, while my primary plan is to stay put for most disasters, there are some that would require me to evacuate.

Even though I may lose my home in that example, the good news—if there is any—is that it is a local disaster affecting me and my community, so finding a place to shelter in a hotel or at a relative's home should be easy.

Likewise, if you find yourself facing a devastating house fire, you too will probably have access to hotels, family, or friends in your community, which would serve as a suitable temporary shelter.

In the event of other local emergencies, such as tornadoes, earthquakes, floods, hurricanes, etc., there will be many more people desperate for a place to stay. It may take time for you to secure shelter if you haven't previously arranged it. In other words, if you haven't prepped.

As I mentioned a moment ago, many of the likely disasters from chapter two may not require you to evacuate at all, depending on where you now live (city, urban, rural). But that decision will be different for each of us.

You should think about this now and have a plan to access shelter for *every* contingency you can think of—go back and look at the list of potential disasters in chapter two.

Thinking about what you will do is a simple exercise that you can do anywhere—in the shower, while driving, or when your mother-in-law is lecturing you. Ask yourself, "where will I go if ____ happens?" for each emergency on the list.

Now, take a moment and think of what your options are for a permanent shelter. These could include your current home or a friend or relative's home a safe distance away. Perhaps you have a second home somewhere that can serve as a safe retreat.

Once you have chosen your options for permanent shelters, ask yourself this question; "what would cause me to leave my permanent shelter?"

When to Stay, when to go

The issue of whether to "bug out" or "bug in" is frequently debated among preppers. While you may want to dig in your heels and decide this in advance, it is important that you remain flexible. The nature of the crisis, population density at your location, your skills, and your level of preparedness will combine to present the best choice for you at the time.

Consider the following examples.

- You live in a city that is being torn apart by violent riots, but riots have yet to spread to other cities. If you have preps and the ability to protect yourself, you're likely best off staying put and securing your doors and windows. Riots are generally short-term and you'll likely incur more risk trying to bug out.
- You live adjacent to forests and just learned that high winds are blowing an uncontrolled wildfire toward you. Monitor the situation closely WHILE you organize your preps and valuables in your car, which should have a full tank of gas. Know multiple evacuation routes that steer you clear of the fire. If the danger appears remotely possible, bug out to a secure location (following the procedures later in this book for evacuation)—and hope that your house escapes damage.

- Something has happened to take down the entire electric grid, and it appears the outage will last at least several weeks, if not more. This means communications are disrupted and you can't be sure what is going on. If you live in a rural area and are prepared, stay put. If you live in the midst of a city—well, I would get out quickly to a pre-chosen location, but you'll need to decide what's right for you. If you live in the suburbs, it's a tougher call. You'll have to assess the situation, your supplies, your neighbors, sanitation options, ability to defend yourself and so on.

Since this book is focused on how to *start prepping* rather than long-term self-sufficiency and survival, let me encourage you to not worry too much about whether to bug out or bug in. Rather, just implement key recommendations in each of the 10 steps in this book to get yourself prepared. Once you achieve that, you'll have the skills, supplies and survival mindset so you can survive comfortably whether you stay or go.

Where to go

If you choose to bug out or must evacuate, where will you go? Try to identify options in three categories: local, regional and far away. This way you'll have an option regardless of the emergency.

Here are a few reasonable ideas for where to go:

- Hotels.
- Campgrounds.
- Your own rural bug-out location.
- Friends.

- Relatives.
- As a last resort, public shelters.

Do you know how to get to your closest hotels and campgrounds? If not, make a point to find out now. Remember to write down phone numbers and addresses in case you need to make a sudden reservation.

Also, instead of thinking merely about a bug-out location, try to establish a reciprocal bug-out relationship with distant friends or relatives. That way you can each rely on the other for secure shelter in the event of an emergency, as well as know what supplies the other has available.

Many preppers choose a rural destination in a low population density area as a target bug-out location. Regardless of where you plan to go, you must be sure the following criteria are met:

- There is a stored food supply.
- There is a potable water supply.
- There is a plan for sanitation.
- You are able to replenish the food and water.
- You have the ability to defend yourself or stay safe at the location.

CHECKPOINT—STAY OR GO

- Make a list of emergencies that would cause you to leave your permanent residence.
- list where you would go and include maps and alternate routes.

- Write a summary of your current state, what you would like it to become and how you will get there.
- **Example:** I will shelter at my house for all disasters other than house fire or any natural disaster that requires evacuation. If I must leave, my first option will be to drive to my parents' house two hours away. The route I will take will be ____. My second option will be to drive to Motel XYZ, located at _____. The route I will take will be ____. If those options are unavailable, my third option is to hook up my camper and drive to the ABC Campground eight hours away at _______.

NON-PERMANENT SHELTER STRUCTURES

Now let's consider a scenario where you don't have access to your permanent shelter. Maybe you are forced to bug out. Or, perhaps you're not at home and want to get home to your preps but, until you do, you need safe shelter.

There are many options for non-permanent shelters and I've broken them into two categories: temporary shelters and mobile shelters.

Temporary Shelter Structures

Temporary shelter structures include various types of tents, such as wall, dome, and pup tents. Smaller shelters allow for better retention of heat produced by the occupants. They're also lighter, less bulky, and easier to carry.

Other temporary shelters include any number of structures that can be created or used from natural elements. These include caves, debris huts, lean-tos, and so on. This is not a book about wilderness survival, but since learning those skills is both fun and important I've included a number of classes and books you may want to explore later in this book.

Sometimes it makes sense to combine shelter structures. An example of this is setting up a dome or pup tent inside a house. The house serves to keep severe elements at bay while the tent within the house can be used as a warm living room or bedroom.

> **WARNING**: When using tents within the house be careful to not have candles or flammable lanterns within the tent. The tent could catch fire and result in fatalities. Also be sure to NOT use kerosene or gas heat sources in confined spaces. Rather, rely on your body heat and insulated sleeping blankets to provide warmth.

Mobile Shelter Structures

Pull-behind campers are an excellent example of a mobile shelter option. Campers can be pre-loaded with many of the survival items you'll need in an emergency, such as stored water and typical camping gear. When a disaster strikes or an evacuation order is given, you only need to hook up and pull out.

Remember though that it's critical to *always* keep your vehicle full of fuel. If you have to bug out it means there's a disaster and gas stations may not be operational.

Your plan may be to drive the camper to a friend or relative's place, or to a pre-designated campground with facilities. Either way, select three options in your planning phase to eliminate the panic of having to figure out where to go in a crisis situation.

Be sure to plan alternate ways to get there since some roads and bridges could be blocked.

Another example of a mobile shelter solution is simply a backpack and tent. You could hike out if you had to, or simply toss the backpack and tent into the car or on the bike and drive to a secure location to set up camp.

On the downside, this choice makes it more challenging with small children or the elderly, and limits the food and supplies you can carry.

CHECKPOINT—NON-PERMANENT SHELTERS

- Make a list of the non-permanent shelter options that are best for you.
- Write a summary of your current state, what you would like it to become and how you will get there.
- **Example:** I will purchase a three-person tent to keep in my car. I will also research learning how to build a shelter in the wild with the hope of trying this by ______.

Now that you have considered all of your shelter options, turn your attention to keeping comfortable.

KEEPING WARM

I've covered many options for shelter, but just because you have shelter doesn't mean you can't freeze. You need a plan to keep all family members and pets warm, since you don't know what time of year an emergency will occur.

Remember—it may be warm at your permanent shelter when a

disaster strikes, but if you need to evacuate to one of your bug-out locations, it may not be warm there.

At my designated permanent shelter, we installed a wood stove as a back-up source of heat, so that is our primary plan for warmth in a power outage. Since we have firewood and are surrounded by hundreds of acres of trees, heat should never be a concern for us.

Even so, we keep a tent inside the house as well as many tarps. Tarps enable us to block off areas we want to heat while preventing heat from escaping to areas that don't require heat, thereby saving precious fuel. These are components of our back-up plan for warmth.

If you don't rely on electricity or wood heat, perhaps you depend on natural gas or propane furnaces. These fuels can generate heat when the power is out, but electric thermostats and fans that push the heated air won't work, unless you have the ability to produce electricity yourself—more on that in chapter six.

If you plan to stay in a permanent location, whether at your house or someone else's, you may have the ability to start a fire, even if it's a campfire in a makeshift fire ring. You'll need a fuel source such as kindling and firewood and, of course, the heat will likely be outside rather than inside a shelter.

On the other hand, if you use a mobile or temporary shelter you still need to stay warm. Here are some items that preppers commonly use to achieve that.

- If a fire is out of the question, a Mr. Heater may be a great option. At 18,000 BTUs, this small device can heat up to 400 square feet. It is fueled by those one-pound propane bottles you see in the sporting goods section of Walmart and similar stores. It can also use the 20-pound propane canisters typically used for gas grills. It's rated safe for indoor use—just be sure to stock up on lots of propane.

- Dyna-Glo makes a portable indoor kerosene unit that heats up to 1,000 square feet. When you have no alternatives it's an instant heat source, but you'll have to store containers of kerosene.
- InstaFire Emergency Fuel comes in five-gallon buckets and is ready to burn. With a quarter cup you can start a fire for signaling, heating, cooking, or to boost morale in a tough situation.
- For less than five bucks, a magnesium fire starter makes it simple to create a fire from twigs, dead branches, dry bark, grass, pine needles, lint, or paper.
- Don't overlook the importance of a great sleeping bag. There are many to choose from, but we settled on the Coleman North Rim Zero Degree Mummy Bag. They can keep you warm inside a tent, with or without a fire.
- Sometimes the simplest solution is the best, so store thermal underwear for all family members. This makes a great base foundation over which to layer clothes. It's a myth that you lose more heat through your head than other parts of the body, but since you'll probably already be wearing pants and a top, wear a hat and good gloves as well when it's cold.
- Instant hand and foot warmers are great to keep on hand in case the power goes out. They have the benefit of being very portable and can provide comfort whether you bug in or bug out.

- Finally, having several wool blankets is a good idea. I purchased *80 percent military wool blankets* on Amazon for less than $25 that may work well for you too.

CHECKPOINT—WARMTH

- Determine your plan to stay warm in your permanent shelter.
- Determine your plan to stay warm in a non-permanent shelter.
- Write a summary of your current state, what you would like it to become and how you will get there.
- **Example**: I will purchase a Mr. Heater for my house. I plan on closing off doors to heat just our kitchen and master bedroom. We will sleep there as a family and I will store at least 20 small propane canisters. In my car I will also store hand warmers and a magnesium fire starter. I will practice starting a fire with it as soon as I get it.

KEEPING COOL

Keeping warm is not always the priority in an emergency situation. If you're on the middle floor of a cramped city apartment building above gridlocked cars in sweltering heat, your family will be equally at risk and you need a plan to stay cool.

Likewise, imagine enduring searing August heat for weeks without air conditioning, courtesy of a Category 5 hurricane. That's exactly what happened to south Floridians in 1992, as Hurricane Andrew left 1.4 million without electricity and left hundreds of thousands homeless.

It can, and will, happen again.

Many regions rely heavily on air conditioning to keep people cool and comfortable. Think—anywhere in the broiling southwestern or oppressive southeastern U.S. for most of the year.

Even northern cities are at great risk during heat waves. In July 1995, a staggering 750 people died from heat exposure over five days in a brutal Chicago heat wave.

In cities the problem is exacerbated since there is essentially no grass or trees and few swimming pools. Instead, steaming pavement, blistering concrete, and suffocating car exhaust can combine with high population density to create unbearable heat without air conditioning. In situations where the power grid is down, it would be a real emergency for millions of people.

Plan now for methods to stay cool should you find yourself in that situation. To get you started, here are 30 strategies to cool down without home air conditioning.

1. Don't forget that you can sit in your car and use its air conditioner to cool off. This is another reason why it's important to always keep your fuel tank filled. Just keep your car parked in the shade with the windows cracked when you're not using it.

2. If you have electricity or a generator, use ceiling or window fans. Just remember that fans cool people but not places, so position yourself to feel the breeze.

3. Use a swimming pool in your neighborhood, but only for the first day or so after the power fails since the pool pump will be inoperable.

4. Seek shade in the woods, which may reduce peak tempera-

tures by 20-40°F.

5. Wear wet sweatbands or bandanas to keep cool.
6. Use a misting bottle.
7. Tint your home windows with darkening film now.
8. Soak a t-shirt in a pool, stream or sink, ring out and put on.
9. Lie down on a wood or concrete floor since they'll be cooler than carpet and heat rises. Basements are ideal, but even crawl spaces are MUCH cooler than the house above them.
10. Build, buy, or make awnings over windows, preferably before any disaster strikes.
11. Install or make shutters, such as plantation shutters.
12. In a two-story house, open windows on the top floor that face away from the wind direction and open windows on a lower floor that face the wind direction. This will allow air to flow and heat to escape upstairs.
13. Open windows at night to let cool air in. Get up and close just before the sun hits your house.

NOTE: In a crisis situation, be aware of the operational security threat of easy entry into your home if the windows remain open, so remain on guard.

14. If you have long hair, tie it up.
15. Eat cooling foods and herbs. Among the most cooling foods are cucumbers, mulberries, and watermelon, and re-

frigerant herbs include lemon grass, hibiscus and the mint family. Make a tea and sit in the shade!

16. Close drapes, shades, or blinds. Buy or make them if you don't have them.
17. Avoid caffeine and alcohol since both dehydrate the body.
18. Avoid using a stove to cook indoors.
19. Eat smaller meals since the larger the meal the more metabolic heat a body produces to digest it.
20. Drink lots of water, but don't drink very cold water. Your body needs to heat the water to core temperature, so doing so can actually make you feel warmer.
21. Place hands and feet in a bowl of cool water or a stream.
22. If you find it hard to sleep at night because it's so hot, wet your hair.
23. Likewise, sleeping in a waterbed will keep you cooler.
24. Sleep in a hammock.
25. Use silk or satin pillowcases and sheets to feel cooler.
26. Light-colored loose clothing is a must. Go nude if you can—hey, here's your chance!
27. Do any physical activity in early morning or late evening and sit in the shade the rest of the day.
28. Eat spicy food! People in hot locations, such as India, Central and South America, Africa, etc. love spicy foods. Why? The hot pepper and spices causes one to sweat, which results in heat loss that cools the body.

29. Locate ponds, lakes, streams, waterfalls and take a dip. (Add a map of lakes, streams and waterfalls to your preps now.)
30. Locate caves and caverns, which will be much cooler.

CHECKPOINT—KEEPING COOL

- List the ways you would keep cool at your permanent location.
- List the ways you would keep cool if you were forced to bug out.
- Write a summary of your current state, what you would like it to become and how you will get there.
- **Example:** I will keep cool by using a ceiling fan if we have electricity. Otherwise, we will open windows at night and seek shade during the day. I need to purchase thermal lined shades for my house so that I can close them during the heat of the day. For non-permanent shelters, I will seek shade during the day and only travel at cooler times. I will keep water and a misting bottle in my car's backpack along with my tent, etc.

Americans believe they can survive an average of 16 days if forced to remain in their homes during a disaster. However, over half (53 percent) of Americans don't have a 3-day supply of nonperishable food and water stored at home.

Adelphi University Center for Health Innovation Poll

CHAPTER FIVE

PREPAREDNESS STEP 3: FOOD STORAGE

Since 1960, the size of the average house in America has more than doubled to over 2,600 square feet. At the same time, the average household size has shrunk from 3.3 to 2.5 persons, so we're not filling that surplus space with children. Rather, we buy more cars and televisions than we need, hoard clothes that no longer fit and fill garage shelves with cans of old paint.

Clearly the average American has the means to purchase stuff and the room to put it. But when it comes to storing food for a rainy day, our houses are virtually empty.

Grandma knew the importance of food security, so she always had a full pantry and an overflowing garden. As a mother of hungry children during The Great Depression, she knew how hard it was to produce food and how much her family depended on it, so she never took it for granted.

All that has changed today, as our relationship with food is very different than ever before. Modern generations have the luxury of believing something that no human has ever believed: that food is a throwaway item to be taken for granted—something that will always be on a grocery shelf down the street or at a restaurant.

A recent Slate Magazine article not only reinforces the widespread belief that supermarket shelves will always be fully stocked, it actually encourages people to *not* stock up.

Written by a meteorologist (of all people), the article was titled *Stop Buying Bulk: It ends up wasting food and money. Shop more frequently instead.*

Slate, owned by the former Washington Post Company, ran the article on June 30, 2015 with a title that clearly encourages people to not stock up on bulk food. Indeed, as the author states bluntly, people should "Just shop more often."

This absurd recommendation steers consumers down exactly the *wrong* path, in my view. Most American households have less than three days of food on hand as it is, and clearly the author believes they should have even less.

In the article, the author claims that American consumers throw away 40 percent of their food purchases, an estimate reported in a 2013 Bloomberg article called *Living in the United States of Food Waste.* But the food waste reported concerned perishable foods, such as fresh vegetables, cold cuts, fresh fruits and soft cheeses, and not bulk grains or canned foods.

The Slate article glibly concludes, "After all, you can't *waste* what you don't buy in the first place."

If a disaster occurs, the meteorologist may be surprised to learn that you also can't *eat* what you don't buy in the first place.

Unfortunately, people read and are influenced by articles such as these, so those who do will have a reason to cling to their normalcy bias and remain unprepared.

Nevertheless, the article does underscore one of the reasons many people don't store excess food: they're afraid it will go to waste. While spoilage is a valid concern with perishable foods, prepping is *not* about storing fresh foods. It's about buying and storing foods in bulk that have a long shelf life.

I believe there are six reasons why we, as a society, don't store enough food to last us through hard times.

1. We don't think we have space to store it.
2. We don't know what to store or how to store it.
3. We eat out much more often, so therefore feel we don't need to store food to cook.
4. We think we can't afford extra food.
5. We're concerned about food spoilage, and most troubling,
6. We believe that food will always be "there."

In this chapter, I'll show you how to create a larder that will last a long time. If disaster strikes, the supermarket shelves may be empty, but your foods will be there to sustain you and your family. But first, let me address a myth regarding the shelf life of food.

The Expiration-Date Myth

The concern about spoilage of fresh foods is understandable, but many consumers seem to have a deep trust of those magical "dates" on canned foods. I'd like to dispel some myths about that right now.

Canned foods are generally "good" far beyond the dates stated. In almost all cases, the dates stated on foods aren't expiration dates anyway; rather, they're "use-by" dates.

The use-by dates on cans and packages serve to protect the *reputation* of the food. They have nothing to do with food safety, as the U.S. Department of Agriculture's own website clearly states:

> "'Use-by' dates refer to best quality and are not safety dates. **Even if the date expires during home storage, a product should be safe, wholesome and of good quality if handled properly**."

Actually, except for infant formula, product dating is *not even required by federal regulations*.

While they may not be required, generally you'll see manufacturers use one of three types of dates, none of which is an expiration date.

- A "Sell-By" date, which simply tells the store how long to display the product for sale.
- A "Best if Used By" date is what the manufacturer recommends for best flavor or quality. It is not a purchase or safety date.
- A "Use-By" date is the last date recommended for the use of the product while it's at peak quality. The manufacturer of the product determines the date.

Of course, manufacturers have an incentive for consumers to purchase more food, so there is a temptation for them to recommend short-term dates to encourage more frequent purchases.

Numerous studies show that foods are viable long after the stamped dates.

In 1983, a fascinating study published in the Journal of Food Science reported on canned food analyzed from the Steamboat Bertrand, which had sunk over 100 years before in 1865.

The findings?

National Food Processors Association (NFPA) chemists detected *no microbial growth* and determined that the foods were as safe to eat as when they were canned. The chemists added that while significant amounts of vitamins C and A were lost, protein levels remained high, and all calcium values "were comparable to today's products."

A prepper's remedy for the loss of vitamins is, of course, to simply store and rotate multi-vitamins in his prepping supplies.

These studies don't surprise me, for proper canning creates a vacuum that prevents microorganisms and air from entering the jar and contaminating the contents. As long as the seal is good, the contents should be good, which is why I'm comfortable eating a jar of stew from my pantry—even if I canned it 20 years ago.

Evidently authorities agree with my view.

In a food safety fact sheet, Utah State University Food Safety Specialist Brian Nummer wrote,

> "For emergency storage, canned foods in metal or jars will remain safe to consume as long as the seal has not been broken."

In another study, NFPA chemists also analyzed a 40-year-old can of corn found in the basement of a home in California. Again, the canning process had kept the corn safe from contaminants and from much nutrient loss. In addition, the chemists said the kernels looked and smelled like recently canned corn.

So as these scientific analyses show, canned foods are an excellent option for preppers; one I'll discuss in detail shortly.

The bigger challenge is to convince people that supermarket shelves can empty FAST when food is most needed, due to the fragility of our food system.

> "**New York City Supermarket Shelves Empty Out,**" was the headline from a January 2015 Washington Post column. The article stated that "meat shelves were all but bare" and reported that nervous, frustrated customers shoved past each other, muttering, "I can't take this!"

Of course we see these headlines anytime there is a predicted blizzard or hurricane, particularly in densely populated areas. The irony is that, when consumers are staring a potential disaster in the face, the first things they grab from the shelves are perishable items such as bread, milk and eggs—foods that often require refrigeration and will sustain them a couple of days rather than weeks or months.

Many people, and you may be one of them, simply assume that they won't wait to the last minute—that they'll stock up the first sign of an approaching storm.

The dilemma is that your food supply is threatened by more than slogging hurricanes and blizzards that can be predicted a week in advance. Any number of crises listed in chapter two, from devastating power outages to a financial crash, could cause not only foods, but also most supplies to be stripped from shelves within hours.

You probably see headlines and read the stories of people desperately seeking food and jobs in Greece, Venezuela, Argentina, Spain and other parts of the world, but—there's a problem. You don't believe *that* can happen here, do you? You figure you'll get a "heads-up"—that you'll see the warning signs far enough in advance.

Your normalcy bias tells you it won't happen to you because it never has in your lifetime.

Stop listening to it. It's trying to get you killed.

I realize that the idea of massive food shortages may seem unrealistic to some of you—perhaps to most of you. Let's not debate whether or not it's likely, since we'll probably agree that it isn't. Instead, just answer this question:

What possible benefit is there to you of *not* storing emergency food and water for your family?

Even your normalcy bias doesn't have a good answer for that, so let's focus on how you can become your own grocery store. After all, you're always going to want to eat. If you store the right foods properly, they won't go to waste.

Let me share with you how we built our deep larder, and why we made the choices we did. Then, I will outline a specific meal list for a two-week emergency food supply to get you started.

THE DEEP LARDER: YOUR GROCERY STORE

I want to encourage you to think about food storage in layers. Many frugal preppers began their food preps by simply purchasing an extra can of tuna or box of pasta during their shopping trips. Later they added cans of beans and chili, bags of rice and other long shelf life foods they enjoy anyway. Over time they found that they had accumulated a deep larder that can sustain their family for weeks or months.

This is a fine way to build a home pantry, but the result generally lacks mobility since the stored foods are bulky and heavy.

Many preppers go far beyond this basic level of preparedness and include bulk and ready-to-eat freeze-dried foods that are lightweight and portable.

Like many prepared people, I've incorporated both of those approaches and have gone much further. We're fortunate to be able to store food every imaginable way—in buckets, pouches, jars, bags, cans, boxes and, since we have the land, on trees, in the garden, on the hoof and swimming in the pond. Regardless of your situation, you can store a supply of food to get you through tough times as well.

Tuna, pasta, grains, and similar items have a long shelf life, measured in many years. Of course, perishable fresh fruits and vegetables have a shelf life measured in days. In between those two categories are foods that you can think of as semi-perishable foods that generally retain quality for several months. These include boxes of cereal, ground coffee, instant foods (noodles, oats, etc.), dried fruit, cake mixes, etc., although you can repackage these items to extend freshness.

A simple introduction to food storage is to try to always have at least one or more back-ups for each semi-perishable item. For example, if you enjoy cereal, make a habit of purchasing three boxes. When you open your second box, let that be the trigger to tell you to purchase two or three more. Do this for your ready-to-eat noodles and similar foods and you'll always have a couple extra of each on the shelf. Be sure to rotate them so you're always eating the one with the closest expiration date.

Now, excluding fresh foods, I recommend five layers of food storage for your deep larder.

Pantry Layer One—Bulk Food Storage

The advantage of bulk foods is that they generally have a very long shelf life and are easy to buy. They include common items such as rice, beans, grains, honey, pasta and more.

We often hear the phrase *food* storage in relation to how to start prepping, but there are other *nutritional* requirements beyond simply "food." So before we get to the actual food, let's start with salt and minerals.

I store LOTS of salt and encourage you to as well. It's cheap and available in 25-pound bags or larger. We expect to use it for flavoring, of course, but also for our animals as well as to attract wildlife in the event of a long-term crisis.

Salt is one of the food storage items you definitely don't have to worry about regarding shelf life. It's been around forever and will last about as long. Be sure to use iodized salt for your consumption in a long-term disaster. Unless you're eating a saltwater diet of fish and kelp, you likely won't get enough iodine.

Curing Salt is something you rarely see on a prepper's list, but we keep plenty of DQ Curing Salt on hand for meat curing. Sometimes called "pink salt," curing salts are a mixture of table salt and sodium nitrite, which inhibits the growth of bacteria, specifically *Clostridium botulinum*. In a grid-down scenario with limited access to medical care, we wouldn't want to battle botulism. Curing salt will ensure our cured meats are safe. Of course, to go along with this you'll need to learn the skill of meat curing. An outstanding book to help you learn that is *Charcuterie*, by Michael Ruhlman.

You may not need curing salt in a short-term emergency, unless your freezer fails and you want the ability to cure meats. However, in that case curing salt and perhaps even a supply of sausage casings would be indispensable.

If you don't want to cure your own meats, consider picking up a salt cured country ham. These uncooked hams are safe stored at room temperature. Because they contain so little water, bacteria can't multiply in them. Heck, even grandma knew that.

While the preservation method makes the shelf life of a country ham practically unlimited, you may want to cycle through them every year or so, since pigs (and hams) today don't have as much fat as in grandma's time. Back then the fat prevented the meat from drying out. With today's lean hogs the hams are still safe to eat when aged longer, but they may become tough and brittle.

Beyond salt, we use and rotate multi-vitamins. Just keep a few bottles of your favorite so you can always have a back up for several months.

And, as you might expect, we store *lots* of bulk foods in our pantry.

These include rice, wheat, oats, honey, sugar, beans, and pasta, and I recommend you do the same. Stored properly, they will last well over a decade—up to 30 years in some instances.

We like purchasing cases of food in #10 cans because their manageable size allows us to take one can out at a time and use it within a month or two. Then we keep the remaining cases and cans on the shelf for back up, meaning we can easily have a year's supply of the most important ingredients.

Even though we like the #10 cans, we often purchase rice, beans, sugar, and grains in 25- or 50-pound bags. Even though I prefer brown rice over white, we choose white rice for long-term storage. Properly stored, white rice can last 25 or 30 years on a pantry shelf. By contrast, brown rice has a limited shelf life measured in months, due to the oil in the bran layer of brown rice that can turn rancid.

I realize bags that large can seem intimidating to some, but there's no need to worry. We simply repack them into manageable sizes when we get home.

For example, we have cleaned and set aside many two-liter bottles over the years. As we purchase large bags of rice we simply pour the rice into these dry bottles and tap them down until they are tightly packed. Before screwing the lid closed we lay an oxygen absorber on top. Then we simply bring one two-liter bottle of rice into our pantry for everyday use while storing the rest on a shelf with our other food preps.

We take the same approach with other bulk foods, sometimes using Mylar bags and five-gallon buckets instead of bottles.

A Mylar bag is a food-grade bag that blocks oxygen, moisture and light, the three big enemies of food storage—you can think of it as a flexible metal can. This is a much better solution than storing food directly in food-grade buckets, because the plastic lining of the bucket is a poor vapor barrier. Over time, moisture and oxygen penetrate, reducing the shelf life of the food.

Mylar bags, combined with oxygen absorbers, add years of shelf life to stored foods. Once we fill the bags with food and oxygen absorbers, we seal the bag with a hot iron and place a Gamma Seal lid on the bucket. The Gamma Seal lid unscrews easily, allowing us to remove contents without having to take the lid off.

In addition to shelves, we often use non-working refrigerators to store our foods, but you may not have the room for that. If not, then I'm sure you can get creative in finding places to store food throughout your home.

The nice thing about repacking bulk food this way is that you can tuck bottles of food in lots of places; under beds, in closets—even under the bathroom sink. Or just toss out the cans of old paint and store them there.

You can get these bulk foods at your favorite warehouse store such as Sam's or Costco, but I encourage you to visit a Mormon LDS cannery and get what you need for MUCH less. While I don't personally subscribe to the Mormon teachings, I appreciate their perspectives on preparedness and the resources they share.

The last I checked there were over 100 LDS Home Storage Centers across the United States and Canada. You can find a current list at tinyurl.com/LDSlocations. These centers allow anyone to pack dry food in #10 cans for long-term storage of 20 years or more.

We have canned food ourselves at LDS canneries and use many of their foods, including flour, beans, non-fat powdered milk, oats, pasta, wheat, and more. I highly recommend this approach for anyone interested not only in long-term food storage, but storing foods that you'll use every day anyway.

I can't calculate how much money we now save on fuel from not having to run to the grocery store each time we hit the bottom of the flour bag. Now, we simply go to our storage room, pull down another can of wheat, flour, rice or whatever we need, and continue cooking.

We store food for convenience more so than in anticipation of a specific disaster.

An added benefit of the #10 cans is that they can be reused for packing other preps. After they're emptied, we clean and use them for fire-starting materials, first-aid supplies, etc. They protect the contents and stack very easily.

If you can't visit an LDS center in person you can order their products online at tinyurl.com/ldspreps.

Other bulk food ingredients we store include jerky, cases of peanut butter, dehydrated nonfat milk, baking powder and soda, yeast, spices, rennet, cheese- and meat-curing cultures, and many gallons of oil and vinegar. We also store honey, which never goes bad, in five-gallon buckets. The cooking oils can go rancid in time, but we repurpose them for soaps or burning in oil lamps as they approach that level of degradation.

Pre-ground flours and meals can go rancid once the oils have been released, but whole grains last much longer. Just do what we do and grind flour as you need it. Keep small amounts ground on hand as you would normally do when purchasing a bag of flour.

In terms of grinding, we have both a manual grain mill and a Wondermill electric grain mill to grind flour from wheat and corn. Our Country Living hand-crank grain mill was expensive but it is very well built. We expect it to be an heirloom that we can pass down to the next generation.

Pantry Layer Two—Freeze-Dried Food Storage

The Inca Empire is credited with discovering the freeze-drying method over 1,500 years ago. Since then, people in the Andes have long used freezing mountain temperatures and a scorching sun to prepare chuño, a freeze-dried potato pulp.

At harvest time, farmers would spread potatoes on the ground and wait three days until they were frozen solid. They then stomped the frozen potatoes with bare feet and allowed the mash to dry in the sun until it darkened, leaving them with a supply of chuño to last the rest of the year.

Modern freeze-drying begins with flash-freezing food. The frozen foods are then placed in a vacuum-tight enclosure and dehydrated under vacuum conditions with careful application of heat. With freeze-drying, the ice turns directly to vapor and is vacuumed out when heat is applied. Freeze-dried foods generally have no added preservatives.

This process ensures that the cell structure of the food remains intact, allowing it to retain the color, shape, flavor, and nutritional value of the food better than air dehydration. Since the cell shape is not changed, freeze-dried foods look more like their fresh counterparts and rehydrate faster compared to those that are air dehydrated.

Essentially, freeze-dried food is made to be rehydrated, which can be accomplished with either cold or hot water.

Many new preppers turn to long-term freeze-dried foods as the first step in storing food. That's a good idea if you have the money and don't have the time or ability to can and dehydrate your own food. If you do have the time, then you can learn some valuable prepping skills by canning and dehydrating your own, but freeze-dried foods definitely have a place due to their numerous advantages.

Freeze-dried foods provide the benefit of an extremely long shelf life, but that shelf life comes at a higher expense. They are very lightweight since so much of the water has been removed, but then again you'll need water to reconstitute them, so plan for that in your water storage.

By contrast, most canned foods have water or juice stored in them, meaning you may be able to store less water.

It's common to see ready-to-eat freeze-dried meals marketed to campers or preppers. These include chili, spaghetti and meatballs, and so on. We do stock many freeze-dried meals, but mainly we rely on freeze-dried *ingredients* from Thrive Life rather than freeze-dried *meals*, because it keeps us in the habit of using them daily.

In terms of purchasing food for long-term storage, we settled on Thrive Life for the reasons I describe on my website, SelfSufficientMan.com. They use superior ingredients, and many are organic that last up to 30 years. For our peace of mind, there's something about seeing those freeze-dried cans on the shelves that puts us at ease, regardless of the latest Ebola scare or stock market crash. If you're interested, you can see the Thrive Life products online at selfsufficient.thrivelife.com.

Most of the time freeze-dried foods come in #10 cans. Thrive Life offers those as well, but we generally purchase the smaller pantry cans. Since we like to make a habit of using these ingredients all the time, having them in smaller cans makes it easy to store them on the shelf with other items we commonly use.

For example, my daughter loves bananas, so we buy Thrive's freeze-dried bananas in the pantry can size, which is only 13 servings compared to the 42 servings in the #10 can. The shelf life of freeze-dried bananas is a whopping 25 years if they're stored properly!

These pantry cans, like all canned foods, can be tucked anywhere. Under sinks and stairwells, on top of your refrigerator, on closet shelves, in empty suitcases, under beds—you get the idea. However, I recommend you avoid attics and non-insulated garages, as temperatures above 70°F will reduce shelf life.

If you're new to freeze-dried foods, buying some small pantry cans is a good way for you to ease into food storage and determine what your family likes. Try a can of fruit and learn how freeze-dried food works. Eat it right out of the can, add to trail mix or rehydrate and add to cereal or oatmeal. You can also include it in baked foods, such as banana bread made with freeze-dried bananas. You'll get a sense of how to use freeze-dried ingredients and what other foods you may want to store.

Start small, find out what you like and, most important, *store what you enjoy eating*. It's nice to have a full pantry, but you'll have enough challenges in a disaster—having to eat unpalatable food needn't be one of them.

In addition to the Thrive single-food ingredients, we also store pouches of freeze-dried meals that are ready to eat. We store some from Thrive, but one of our favorites is the Mountain House brand spaghetti with meat sauce. If you have some picky young eaters, this will likely satisfy them.

These pouches are also great for camping trips or to carry in survival bags. Try several to see what you like. They may not be quite as good as grandma's home cooking (what is?), but in an emergency, they're a fantastic resource.

Even though they are convenient, we don't use these pouches of ready-to-eat meals too often. Rather, we store them to use on camping trips and to be secure in the knowledge that we have meals to sustain us at home and on the road if disaster strikes.

Mountain House has lots of great freeze-dried meals and you can find them all at tinyurl.com/mhmeals.

Pantry Layer Three—Canned Food

Each layer of the pantry has its own pros and cons. With canned food, benefits include universal availability and broad selection. Whereas the selection of dehydrated, bulk or freeze-dried foods is limited in supermarkets, there are many aisles devoted to canned meats, fruits, vegetables, nuts, drinks and the like.

We can thank the French military for canned foods. In the late 1700s, soldiers fighting in Napoleon's army suffered frequent outbreaks of scurvy on a diet of poorly cured meats. The French government offered a 12,000-franc reward to the person who developed a safe and dependable food-preservation process.

In 1810, after 15 years of experimentation, Nicolas Appert, a French chemist, claimed the prize. Appert placed food in glass jars, sealed them with cork and wax, then placed them in boiling water. He observed that food heated in these sealed containers remained preserved as long as the container remained unopened or the seal did not leak.

Today, canned food is an important element of a deep larder, but I have purposely listed it as the third layer, behind bulk storage and freeze-dried food.

We prioritize those two methods for long shelf life and the option to cook up pretty much anything. Of course, relying on the first two layers of pantry storage does require one to be able to cook, so I hope you too enjoy it.

Managing canned foods requires more effort than dealing with dehydrated or freeze-dried foods. The cans are heavy and they take space, although you can tuck cans here and there if you don't have a dedicated space.

Plan on stocking whole meals in a can, such as beef stew and chili in addition to cans of fruit, vegetables, and pasta sauces.

Shelf life is normally stated at one or two years, other than canned meats such as salmon, which are often listed as five years or longer. But read what I wrote earlier about the myth of expiration dates—properly canned foods won't go "bad" when they hit their stated dates. They just *may* not retain their optimal quality as defined by the manufacturer.

My rules for stocking up on canned foods are

1. buy what you like to eat,
2. buy what you need,
3. stock a variety of foods (meats, vegetables, fruits),
4. stock up when they're on sale or buy-one-get-one free and
5. rotate your supply.

The "buy what you like" recommendation is common and obvious, but let me explain the "buy what you need," for I'm not *just* talking about what nutrition you need.

Let's say you're stocking a bug-out vehicle or retreat where access to water may be a challenge for you. In that case you can do yourself a favor by choosing canned foods that have a lot of water.

Do you ever debate over whether you should buy the canned tuna or sardines packed in water or—say—mustard? Opt for the water, and you'll have at least a little something to drink. If sodium is a concern for you then be sure to look for "no salt added" or "low sodium" canned foods.

In particular, canned vegetables tend to have a lot of water. For instance, the USDA requires short-cut green beans in an 8-ounce container to weigh at least 4.5 ounces, filling about 56 percent of the can. This leaves 3.5 ounces, or 44 percent, for water that you can drink.

The good news is that one to two years is a long time—certainly longer than pretty much any emergency you're likely to face. So by rotating your foods you'll be able to amass an impressive food pantry.

As for us, we do have some purchased canned foods, such as tuna and salmon, but we can as much of our own food as possible. Our preference is to grow what we eat.

To do that we use the All American Pressure Canner and I highly recommend these over other brands. One of the undeniable facts of preparedness is that certain tools are relatively costly, and these canners are no exception. Then again, some people spend their hard-earned money on golf, travel, or gourmet dinners, so we all have choices when it comes to priorities.

In addition to pressure canners, we also have water bath canners for high-acid foods such as jams, jellies, preserves, fruit, and relish. However, meats are low-acid foods that require a high-heat preservation method.

Low-acid foods include vegetables, tomato products with added vegetables or meat, meat and game, soups, stews, seafood, and poultry.

All low-acid foods must be processed at a temperature higher than that of boiling water, which means pressure canning. The higher temperatures are required to destroy naturally occurring spores that can cause botulism.

For me, a high quality pressure canner is a *very* high priority. These canners are indispensable, will likely last your lifetime and someone else's and will ensure that you practice food preservation skills.

Of course, canners are of little use without the canning jars, so be sure to store plenty of Mason Jars, rings, and lids.

When I wanted to stock up on Mason Jars several years ago I asked my local feed store to order me an entire pallet, or approximately 1,000 jars. They did, but it took me many months to convince them that I *really* wanted them for preserving food and not for what they suspected—to make moonshine.

Luckily, you can often find inexpensive used cans and rings at thrift stores and yard sales. Just be careful to check for cracks and chips.

You must also know how to use them, so if you don't have home canning skills, don't wait until the last minute to learn. Start practicing now!

Another benefit of having a canner is that often, during disasters, the power goes out and food in freezers can spoil. A pressure canner allows you to can those meats and vegetables so they'll not only not spoil, but will be good to eat for many, many years.

One of my favorite methods with a pressure canner is to can steaks or roasts that are nearing the freezer burn state. By cutting them into chunks and adding vegetables to the jars before I pressure can them, I have jars of tender pot roast and vegetables on the shelf waiting to be consumed.

We pressure can meatloaf, green beans, meatballs, ground beef, carrots, turkey, squash, chicken, venison, corn, rabbit—even bacon.

With a pressure canner, you'll never have to throw out meat again because it's "going bad."

Pantry Layer Four—Dehydrated Food

Dehydration is one of the oldest methods of preserving food, dating back some 14,000 years. Our earliest ancestors used the sun and air to dry out and preserve food. That approach still works, but today we commonly use a slow air-drying process.

Dehydrated foods differ from freeze-dried foods in a couple of key ways.

For one, the cells of dehydrated foods do change shape and reduce in size, which is why they take longer to rehydrate than freeze-dried foods.

Also, many purchased dehydrated foods contain sugar, salt and preservatives, whereas freeze-dried foods do not.

A third and significant difference is that it's easy to inexpensively dehydrate food yourself at home. Harvest Right does make a home freeze dryer, but it costs approximately $4,000 as of this writing. By comparison, dehydrators cost less than one tenth of that.

For years, we used an inexpensive round dehydrator, but it produced inconsistent results. Now we use an Excalibur dehydrator, which performs much better. This is largely due to the Excalibur's fan and heating element being located in the rear, which allows for even air and heat distribution.

By contrast, the round dehydrators have the fan on the bottom, so the bottom layer dries much more quickly than the top layer. In our experience, food retains its color much better and the dehydration quality is superior in the Excalibur.

Dehydrated foods form a very valuable food pantry layer, but many people aren't sure how to use them. As for us, we use our dehydrator to make jerky, for sure, but we have found so many practical uses since we have a garden and homestead.

For instance, during the long summer months when the hens are laying, we get far more eggs than we can consume. So we dehydrate and vacuum seal them in Mason jars to last through the winter months.

When the strawberries and fruits come in, we use the fruit leather trays on our dehydrator to make preservative-free fruit leathers. We also make leathers from tomatoes, which can then be dehydrated for pizza or spaghetti sauce, or mixed with dehydrated carrot, celery, and onion to make a soup stock.

You can also do this even if you don't homestead. Just buy food when it is on sale or pick your own at local farms.

While I recommend dehydrating your own food, as we do, I recognize that many readers will prefer a more convenient alternative.

If that describes you, simply start by purchasing some dehydrated food to see what you like. Of course, raisins are dehydrated grapes and I suspect you've enjoyed your fair share of those. But venture out with other dehydrated fruits, vegetables and even meats. After all, jerky is dehydrated meat.

Add dehydrated fruits to your oatmeal, to recipes for cakes and baked goods, or simply snack on them right out of the bag. You'll probably find that you enjoy them, but that there is a limited variety in the grocery store. That's when you'll want to start dehydrating your own.

Pantry Layer Five—Frozen Food

Many people store food in the freezer, but there are several downsides to relying solely on that approach. For one, a freezer takes a lot of space. For another, it takes electricity and appliances break. Also, some food spoils in the freezer after only a few months or a year at most.

By contrast, these negatives do not apply to canned, freeze-dried or dehydrated food.

Typically we use a freezer as a back-up source for more perishable items. When we bake breads or cakes, we devote one baking day to making many loaves and freeze them all. That way when we want to use one we simply take it from the freezer. You can take the same approach when you see bread on sale.

When butter is on sale, we buy lots and freeze what we don't use right away. If we're cooking a dish such as chicken pot pie, we'll make two at the same time and freeze one.

The idea is that, for the most part, we use the freezer to store foods that need to be consumed within a few months. The exception is any large roasts or steaks that we may store, but even in that case we pressure can and store more meats on the shelf than we do in the freezer.

Due to their energy requirements, freezers are my least preferred method of preserving food. On the other hand, I do enjoy ice and it's hard to get that from a can, so I certainly appreciate freezers.

If we found ourselves without electricity for a long time and our frozen foods were in jeopardy, we would quickly can what we could on a portable gas stove or over an open fire. We would use the water bath method to can the fruits, since they are high-acid foods, and we would pressure-can the vegetables and meats, since they are low-acid foods. We freeze lard and fats as well, but would likely convert them to soap if our freezer failed. We even can butter in the form of ghee and keep it on the counter, since it is shelf-stable.

Later in this chapter is a section called Power Outages & Food Safety. It describes what we do when the power goes out.

Now that you understand the five layers of food storage, let's stop for a moment.

CHECKPOINT—FOOD STORAGE

- Determine which foods in each category you want to store.
- Identify spaces in your home for them. Where can you put cans, bottles and packages of non-perishable food?
- Write a summary of your current state, what you would like it to become and how you will get there.
- **Example**: I want to store 50 pounds of rice, 25 pounds of salt and 25 pounds of sugar. I can store them in their original bags under my bed. I will order #10 cans of pasta, beans, and oats to keep on garage shelves. I will try some freeze-dried strawberries and add them to my oatmeal this month to see how we like them, and I'll try another fruit next month. Also, I will make room in my kitchen cabinet for 50 cans of food, where I will store 10 cans each of Peas and carrots, chili, green beans, corn, and beef stew at all

times and rotate them for everyday cooking. Finally, I will keep a backpack full of freeze-dried ready-to-eat Mountain House meals in the coat closet and use them for dinner at least once every other month, replacing them as I do so.

HOW MUCH FOOD TO STORE?

You often hear people recommend storing food, sometimes two weeks' worth or more. But *how much* food is that?

I suggest you not think about putting two weeks' worth of food away, but rather concentrate on putting two weeks' worth of *meals* away. After all, that's what your family eats, right?

Let me show you how I would construct a theoretical 14-day food menu for *my* family, which includes two adults and one young child. Even though it requires 42 *meals* to feed three people breakfast for 14 days, it doesn't necessarily require 42 *servings*, since it's reasonable that a young child's serving is half that of an adult. Therefore, in our case we only need 35 breakfast *servings* rather than 42, which is 2 ½ people X 14 days. The same is true for lunch and dinner.

We'll plan a 14-day menu of 35 servings for each breakfast, lunch and dinner. The plan assumes we have water stored and a basic ability to cook, but that refrigeration will not be available.

In planning the menu we'll be careful in counting the servings on a package because we will have no ability to refrigerate leftovers. As a result, we'll make food choices that allow us to cook only as much as we need per meal without any food waste.

Before we get started, let's get the mandatory disclaimer out of the way.

Disclaimer

The contents of this book and this chapter are *not* recommendations for you or anyone else. I'm not an expert on what should be included in *your* food storage. Only you can decide what is best for your family. Allergies, medical requirements, infant and elderly needs vary from family to family, so assemble what is right for your specific needs.

14 DAY EMERGENCY FOOD STORAGE PLAN—TWO ADULTS & ONE CHILD

Staples & Essentials to Store

- Sugar.
- Salt and spices.
- Full jar of olive oil.
- Measuring cups.
- Cooking utensils.
- Manual can opener.

Breakfast For 14 Days

Okay, we need 35 breakfast servings.

Storing **one** #10 can of Quick Oats along with **two** quarts of honey and a little salt will provide *30 breakfast meals*. Since we are a family of three, this gives us *breakfast for 10 days*. We can make only what we need for each meal, so there will be no leftovers to refrigerate.

Let's add **four** 15-ounce cans of corned beef hash and **one** pint of powdered eggs to our pantry. Using one can of hash per meal along with a few tablespoons of rehydrated eggs to scramble enables us to make 12 meals, which is breakfast for *four days*. We dehydrate eggs ourselves, but you can purchase a can of long-lasting scrambled eggs from Thrive. This uses very little water and scrambled eggs help you forget that the power is out.

Now that we have our 14 days of breakfast taken care of, we need to remember to store breakfast drinks. I suggest one large jar of instant coffee, one can of powdered milk, and one large jar of Tang. The milk can be reconstituted to provide an additional beverage choice or blended with coffee to replace creamer.

Simply storing the oatmeal, powdered eggs and corned beef hash provides all the breakfast meals needed for two weeks, but if you're looking for variety, consider adding packages of pancake and muffin mix.

An alternative for "simple" breakfasts is to store a Mountain House breakfast bucket of freeze-dried meals. Buy, store, and forget. Just be sure to have plenty of water stored. Storing freeze-dried meals is a good idea in any event. After all, you may need to evacuate and take food with you.

Lunch For 14 Days

Store **six** quarts of chili or beef stew. Also, store **two** pantry cans of Thrive Golden cornbread mix. Unlike Jiffy cornbread mix, which requires milk and eggs, Thrive requires only water. This will provide enough lunches for *six days* assuming, A) one quart will feed all three people with no leftovers to refrigerate, and, B) each cornbread mix will make enough for three days, with leftovers being wrapped in foil and stored at room temperature.

Store **one** #10 can of elbow macaroni from Thrive or LDS along with **eight** pints of pasta sauce. It may be easy to find sauces in large quart jars, but smaller jars are a better choice for emergency prepping since refrigeration may not be available. Cooking just enough macaroni each meal to accompany a pint of pasta sauce provides lunch for *eight days* with no leftovers. Each #10 can of macaroni has an eight-year shelf life and includes 25 servings, providing more than enough pasta to cure those hunger pangs.

Wash down your meals with water, coffee, Tang, Kool-Aid, or powdered lemonade. Just stock your preference.

Dinner For 14 Days

There are many options for storing shelf-stable foods to create excellent dinners.

Store some **white rice**—at least **five pounds**. You can buy #10 cans from Thrive or LDS, but we normally just buy large bags at wholesale clubs. Then we repackage them in recycled gallon jugs, two-liter bottles or five-gallon buckets to keep the rice safe from rodents and insects.

Also, store about **five pounds** dried beans, **one** can of Thrive freeze-dried corn and a container of chili powder. Cook just enough rice, beans, and corn for the dinner, and toss in some chili powder and salt for a complete meal. You'll have enough dinners for *seven days*. If you'd like, stock up on extra cans of Thrive Golden cornbread mix, but it's not required. Then again, as grandma might have said, "it's hard to be depressed when you have a mess of cornbread on the table."

Now let's plan the other seven dinners using the rice that you already have on hand.

Store **seven** cans of peas, **seven** cans of cream of mushroom soup, and **one** #10 can of Thrive freeze-dried chicken. This allows you to make a casserole to feed all three people each night for seven days. Just make as much rice and chicken as needed for each night, use **one** can of peas and **one** can of soup. That way, there will be no waste, no leftovers. It's simple to do this in a cast iron skillet or Dutch oven and cook over an open fire or on a gas grill.

You have the same beverage choices as before, but remember that wine not only stores well, many bottles improve with age if they're stored properly. So prepping is your sanctioned excuse to build that wine cellar you've always dreamed of.

Morale Improvement

The menus above provide plenty of tasty nutrition to get through 14 days without having to leave the house. I don't know about you, though, but we like to have some goodies to brighten our days. Here are some easy-to-store items that will make a power outage not only tolerable, but perhaps seem like a vacation.

- Store a jar of popcorn. Pop when desired and serve with salt.
- The Mountain House freeze-dried apple crisp is fantastic!
- Canned fruit is a nice treat when you can't get fresh fruit.
- The Thrive Fudge Brownie mix is also fantastic and only requires water. Yum when you need it most.

- Keep a bag of hard candy with your emergency preps and don't touch it until disaster strikes—not even if you're depressed.
- Dried fruit such as banana chips store well. We repack ours by vacuum sealing them.

Once you have this 14-day supply of food assembled, place it on a dedicated shelf and *Don't Use It!* Just keep an eye on expiration dates and replace items when necessary.

Here's a picture of our 14-day supply so you can see how much space it takes—not too much really.

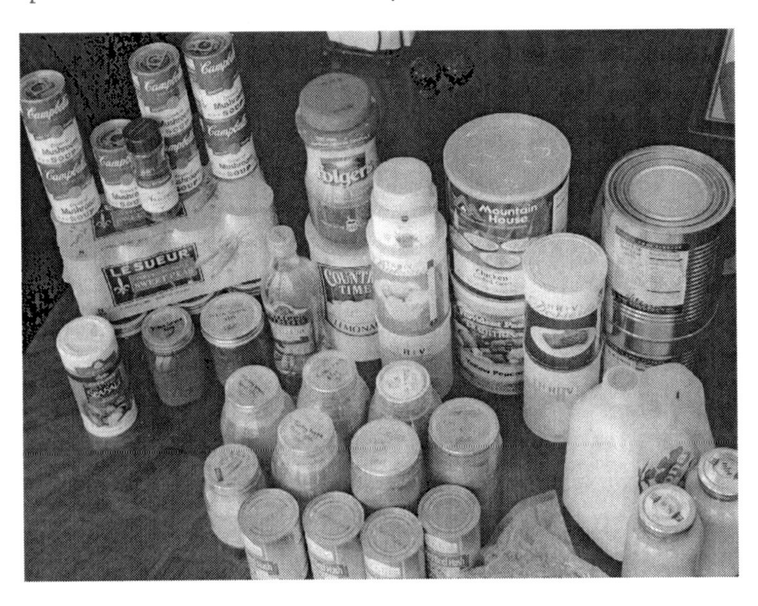

14 Days Worth of Food

POWER OUTAGES & FOOD SAFETY

Now let's turn our attention to food safety. If the power is out for less than four hours, then the food in your refrigerator and freezer should be safe to consume. While the power is out, keep the refrigerator and freezer doors closed as much as possible to keep food cold for longer.

Try to keep your freezer full at all times. If you don't have enough food to fill it, fill the extra space with blocks of ice.

In one of our freezers we simply filled a five-gallon bucket with water, placed it in the freezer and let it freeze solid. If the power goes out, that block of ice will help keep the freezer cold for days. Since a block of ice has far less surface area exposed than a bag of cubed ice, it will stay frozen much longer.

If the power is out for longer than four hours, follow these guidelines:

For the Freezer section: A freezer that is half full should hold food safely for up to 24 hours. A full freezer should hold food safely for 48 hours if you don't open the door, so keep your freezer filled.

For the Refrigerated section: Pack milk, other dairy products, meat, fish, eggs, gravy, and spoilable leftovers into a cooler surrounded by ice bottles, as I describe in the next chapter. Inexpensive Styrofoam coolers are fine for this purpose.

Use a food thermometer to check the temperature of your food right before you cook or eat it. I recommend the Weber 6492 Insta-Read Thermometer.

A common recommendation is to throw out any refrigerated food that has been at a temperature of more than 40°F for two hours. When in doubt, you should follow that recommendation, but there are several foods that I personally would not discard above 40°F, because I believe they are still safe.

I'm not advising you to do this. Rather, I'm sharing with you what I do.

Here are the foods I would keep even if they were above 40°F but lower than 65°F, unless otherwise noted:

- Hard cheeses, such as cheddar, Swiss, Gouda, Gruyere, and Parmesan that are normally aged in "cheese caves" at 52-60°F anyway.
- Butter and eggs, which are commonly stored on the countertop in many countries.
- Fruit juices or opened canned fruits, due to their acidity, although they should not be stored in the open cans.
- Raw vegetables.
- Fresh mushrooms, herbs, and spices.
- Bread, rolls, cakes, muffins, etc.
- I would not discard mayonnaise or tartar sauce unless it was held above 50°F for eight hours or more.

In the next chapter I will share ways to keep food cool when the power fails.

> **NOTE:** It is *not recommended* to put frozen food in the snow! Frozen food can thaw when exposed to the sun's rays even if it is very cold.

In case you're wondering, you may refreeze the food in a freezer if it thaws or partially thaws. Food may be safely refrozen if the food still contains ice crystals *or* is at 40 °F or below. It is up to you to evaluate each item separately and be sure to *discard items* in either the freezer or the refrigerator that have come into *contact with raw meat juices*.

According to the USDA, partial thawing and refreezing may reduce the quality of some food, but the food will remain safe to eat. See the charts in Appendix II for specific recommendations.

FINAL THOUGHTS ON FOOD STORAGE

A 14-day supply is a great start for your food preps, but now that you see how to craft your menu you should set a goal to achieve a 30-day supply.

Even when you hit that goal, many preppers will argue that a 30-day supply is woefully inadequate. As grandma knew, you can *always* be a little more prepared. But how many examples can you or anyone else point to in your lifetime where a 30-day supply would have been insufficient?

Realistically, you're most likely to appreciate having long-term food storage not in the event of common widespread emergencies, such as hurricanes, but rather during personal ones, such as lengthy bouts with unemployment or illness.

I do agree with die-hard preppers that your family will be much more secure with a three-month supply or more of *all* preps, including food, water, money, medical supplies and so on.

However, you can't get to three months without first reaching one month, and you can't get to one month without hitting the two-week mark. So hit that target and go from there.

Actually, if you look closely at the menu you'll see that storing a 30-day supply of food takes up very little space. So no matter where you live, you can likely find room for it.

Of course you'll need a way to cook most of these meals, and I'll cover off-grid cooking methods in the next chapter.

We forget just how painfully dim the world was before electricity. A good candle provides barely a hundredth of the illumination of a single 100-watt light bulb.

Bill Bryson

CHAPTER SIX

PREPAREDNESS STEP 4: ENERGY

Grandma wouldn't have noticed if the electric grid went out back in the "old house." Of course, that's because she didn't depend on electricity, but that doesn't mean that running her homestead didn't require energy. It just means that she and others had to supply the energy, most often in the form of manual labor, wood, or oil.

When she washed clothes by the creek, it was her back and arms that fueled the cleaning. When she sewed or mended a dress, her legs provided the energy to power the sewing machine. If she lost track of time and sewed into the night, oil or kerosene lit the way.

In a power outage, grandma's food wouldn't have spoiled. The rainbow of canned garden vegetables on the pantry shelves needed no refrigeration. As for milk, she simply stored it in the cow until she needed it for butter, or on special occasions, ice cream. And, due to its low moisture content, bacteria couldn't grow in that country ham grandma had hanging from the rafters, so the meat stayed at room temperature—sometimes for years.

But we're not grandma, are we?

Lucky for us, I guess. Whereas grandma had to slave away to produce energy, we mindlessly plug cords into walls and tap a magic stream of electricity. Our children ask us where the electricity comes from, and the best we can do is to point to the power lines—we simply don't know. From powering refrigerators and washing machines to energy-zapping microwaves, DVRs, vacuum cleaners and air conditioners, we are addicted to electricity.

Withdrawal, should it be forced upon us, will be excruciating.

Most people have only a couple of cans of food on the shelves, and few have a pantry at all—stocked or otherwise. The majority of people rely on freezers to store meats, ice cream, and pizza, and refrigerators to chill milk, yogurt, and leftovers.

In other words, the average person's food storage is almost completely dependent on a working power grid.

An electrical grid failure may have had no effect on grandma, but it will crash your life to a sudden halt. For those not prepared, here's what to expect, at a minimum:

- All frozen and refrigerated food will spoil quickly.
- There will be no electric heat or air conditioning.
- Water faucets will not work.
- Internet, television, radios, game consoles, and computers will not work.
- Electric lighting will not work.
- Home security systems and smoke detectors will fail after the battery back-up is drained.
- Neither washing machines nor dryers will work.
- Showers and toilets will not work since water cannot be pumped into the house.

That's the minimum of what you'll lose, and only if you're suffering a *local* power outage.

During widespread outages, expect that you won't be able to refuel your car, have cell phone service or access to emergency help. You also won't have access to ATMs or online banking.

I suspect I know what you're thinking. It's scary all right, but power never goes out for *that* long. How likely is a large-scale power outage anyway?

Well, it has happened before.

The biggest blackout in U.S. history occurred on August 14, 2003, and left 55 *million* people without power in the northeastern U.S.

The cause? A simple software bug in the alarm system.

Actually, lengthy and widespread electrical disruptions are quite common. We just don't recall them unless we personally experienced the hardship they cause.

- Superstorm Sandy left 8 *million* people without power for days in the midst of frigid temperatures.
- Similarly, Hurricane Irene caused over 5 million power outages in 2011.
- In 2012, an intense heat spell and line of thunderstorms left 12 dead and *millions* without power for days in the eastern U.S.
- A 2009 ice storm left three quarters of a *million* without power in Kentucky and Indiana. Three *weeks* later, thousands were still without electricity.

- In 2008, a large ice storm took down power lines from Maine to Pennsylvania. One and a half million people lost power and it took *two weeks* to restore power to everyone.

But those are natural disasters, and the power is always restored—right?

So far, but some people are concerned that our electrical grid is vulnerable to an attack from a high-altitude electromagnetic pulse (EMP).

The fear is that a single nuclear weapon exploded at high altitude above the U.S. will unleash an EMP that would shut down the U.S. power grid and, along with it, communications, water supplies and food transportation.

If the effect is long-lasting enough, it also could trigger a social collapse that could conceivably cause the deaths of tens of millions of people and, temporarily, push the nation back 100 years.

Sound farfetched? These are not the claims of a ranting lunatic. These are the findings of a U.S. congressional commission as reported to the Senate.

That commission was the EMP Commission, formed in 2001. The *Report of the Commission to Assess the Threat to the United States from Electromagnetic Pulse Attack* was released in 2004. Ironically, it was released on the same day as the widely anticipated 9/11 Commission report—which garnered all the news coverage.

A second EMP report was released in 2008 and the findings were just as alarming, as it concluded that

> "The high-altitude nuclear-weapon-generated EMP is one of a small number of threats that has

the potential to hold our society seriously at risk."

The scary part of an EMP scenario is that it only requires one nuclear weapon, detonated 300 miles above the middle of the U.S. Just a single bomb, which could be launched from a small container ship or a freighter parked a few miles off the U.S. coast, could take down the U.S. electric grid for years.

The instant it is launched, the war is already over and we all lose.

Many who fear a nuclear EMP attack point to North Korea and Iran as threats, often suggesting that their test-fire "mishaps" halfway into their projected flight are not the "failures" the media report them to be.

To others, the notion of an EMP seems farfetched, as there is little evidence that terrorists could succeed.

Then again, prior to September 11, 2001, we had no "evidence" that terrorists could hijack and use jet airplanes to attack the crowded skyscrapers and the pentagon. Still, it happened, and thousands lost their lives that day.

Fortunately, the U.S. as a nation survived 9/11, but the effects of an EMP attack would likely be far worse than we can easily imagine.

One person who imagined it for us is William Fortschen, author of the chilling book *One Second After*. In preparation to write the book, Fortschen spent four years researching EMP, interviewing scores of personnel from Congressmen and Generals to local law enforcement. A staggering projection from Fortschen's work is that 90 percent of the U.S. population could die within 12 months of an EMP attack.

Peter Pry, a former CIA officer and head of a congressional advisory board on national security, agrees.

> "I think we're running out of time. If a nuclear EMP happens this gets translated into mass fatalities, because our modern civilization can't feed, transport, or provide law and order without electricity."

Long-time military analyst John Pike, however, was skeptical about this threat, saying,

> "It is just very difficult to imagine how terrorists are going to be able to lay hands on a nuclear-tipped missile, launch it and reprogram it in such a way that it would be a high-altitude burst like that."

What's really scary is that a nuclear device isn't even necessary to deliver our society back to the Dark Ages. The sun can do it for us just as easily, and there's not much we can do about that.

In fact, even those who dismiss the nuclear EMP threat as unlikely acknowledge that the same result could come from a solar storm.

> "It is *virtually guaranteed* that a powerful geomagnetic storm, capable of knocking out a significant section of the U.S. electrical grid, will occur within the next few decades," said Yousaf Butt, a nuclear physicist with the Federation of American Scientists. "In fact, this may well happen within the next few years."

We don't tend to think of outer space when we think of natural disasters, but severe space weather can deliver a far more devastating punch to our modern electrically dependent lifestyles. And according to experts, it's inevitable that it will happen again.

The largest solar storm on record occurred in 1859. Dubbed the "Carrington Event," after British astronomer Richard Carrington, the geomagnetic disturbances were so strong that U.S. telegraph operators reported sparks leaping from their equipment. Some equipment actually burst into flames.

A 2013 report by London-based insurance market Lloyd's of London considered a Carrington-level solar storm to be "*almost inevitable* in the future." The report also estimates that between 20 million and 40 million people in the U.S. are at risk of an extended power outage *lasting up to two years* due to the damage caused by a Carrington-level geomagnetic storm.

Beyond nuclear EMPs and solar storms, there is the increasing risk that terrorists will attack our fragile power grid.

A USA TODAY study found that the nation's power grid is struck by a cyber or physical attack once every four days. The study revealed that transformers and other critical equipment often sit in plain view, protected only by chain-link fencing.

Of course, the cause of a grid failure doesn't have to be a freak doomsday scenario such as an EMP. It also needn't be something as bizarre as a software bug in an alarm system. Most often, it's simply Mother Nature tossing a storm our way or crippling us with a heat wave that creates rolling blackouts.

Maybe she'll just shove over a few trees on your power lines or cake them in ice.

Whatever the cause, here are the facts: we're becoming more addicted to electricity each year, and power failures happen with regularity. Being prepared for extended outages just makes sense.

So, what should you do?

Here are my recommendations for you to achieve basic energy preparedness. This will allow you to get by comfortably for a few weeks without the power grid.

How To Generate Your Own Electricity

In an emergency, you'll want to focus on staying comfortable and survive short-term grid failures—two to four weeks. To achieve this, think first about conserving all energy and powering only the necessities. Freezer, yes. Xbox, no.

One of the first actions is to prioritize needs for energy. For instance, powering a freezer is more important than powering a refrigerator or electric lights, for reasons I will soon explain.

On the other hand, while the contents of a freezer are valuable, in a flooding situation it may be best to prioritize a sump pump over a freezer.

Then again, if you have a well and can use power to provide running water to your home, that may be a pretty important need. Every person's situation will be different depending on the disaster and their needs.

While there are many ways to produce electricity, I'll cover two common methods that will allow you to achieve a *solid* level of preparedness. These methods involve generators and inverters.

You can choose one or the other, or both, but your choice should be driven by what you *need* to power when you are without electricity.

Suppose all you want to do is run the sump pump or keep your food refrigerated during short-term outages. In that case you can get by with a small generator.

However, you may decide you want to provide power to a well pump, computers, Internet access, television, heat and air conditioning, lights and more items. You may need a whole-house generator for that and, depending on how long you are preparing for, you'll need access to a lot of fuel.

Make a list of the appliances you *need* and check both the starting (peak) watts and continuous-load wattage for each. Once you know that, you'll be able to choose the option that meets your needs.

You may want to purchase a Kill-A-Watt. For about $20, this simple device connects to any electrical device and instantly displays the wattage the device is drawing. This way you'll know *exactly* how much power you need rather than what the appliance label says.

But before we discuss the merits of generators and inverters, let's pause for a moment so you can assess your electrical priorities.

CHECKPOINT—ELECTRICAL PRIORITIES

- Think of all the electrical appliances you use.
- Prioritize their importance from most important to least important.
- Ask yourself if certain disasters would change that priority order.
- Write a summary of your current state, what you would like it to become and how you will get there.

- **Example**: I bought a Kill-A-Watt and now know how much wattage my appliances require. Therefore, my top three most important appliances are, 1) well pump, 2) freezer, and 3) window air conditioner. However, if a disaster occurs in cooler months then my third priority would be a small microwave rather than an air conditioner. If I have excess power I will allocate it to Internet access.

Now, let's review the energy production options available to you.

Method One: Generators—Your options for generators include whole-house generators that can automatically kick on in a power outage, or portable generators that you must position and start when needed.

There are numerous advantages to having a whole-house generator, also known as a standby generator. The key advantage is that it delivers much more wattage, often enough to power an entire home.

Standby generators are usually professionally installed in a permanent location with an automatic transfer switch. Within 45 seconds of a power outage the standby generator will start and power lights, refrigerator, TV, and whatever was on before the power loss. Most models automatically run weekly self-tests to ensure they work properly. When electricity returns, the generator will automatically switch back to the utility power source.

Unlike portable generators, extension cords are not required since standby generators connect directly to the electrical panel. This makes it much more convenient to operate appliances normally. Most whole house generators are fueled by either natural gas or propane. If you keep a large propane tank filled just for this purpose, you won't have to worry about storing and rotating gasoline in cans.

The downside to whole-house generators is the cost to purchase and install, which, at a minimum, will run several thousand dollars. They also require installation by a licensed electrician. Of course, another downside is that this option is not available if you rent or live in an apartment.

By contrast the chief virtue of a portable generator lies in its mobility. On a homestead such as mine, I can wheel one to the well house to power the well pump. Likewise, I can take it to the workshop, the house, or anywhere else we need electricity.

Another benefit is that portable generators cost far less than whole-house generators. However, they don't produce enough electricity for heavy drains such as central air conditioning.

Whereas whole-house generators can be connected to either natural gas lines or large propane storage tanks, portable generators have small fuel tanks. If the power outage continued for a week, you'd need about 140 gallons (that's 28 five-gallon containers) of gas to run a typical generator continuously. Just remember that the greater the load on a generator, the more fuel it consumes, which is why it's critical to decide in advance what you *must* provide power to.

With portable generators, you'll most likely have to store fuel in cans, which can be hazardous, and many gas stations will be closed during widespread power outages.

Finally, you'll have to pull a cord to start most portable generators, which may be a challenge.

If you opt for a portable generator, having a tri-fuel generator that can use gasoline, natural gas, or propane will provide the most flexibility. Having a large propane tank allocated as a fuel source for your generator will eliminate the need to store gas cans.

Of course, as is the case with any small engine, portable generators require maintenance, such as oil and filter changes. Unlike standby generators that have a weekly self-test feature, portable generators must be manually started often to ensure they are in good working order.

A final but crucial consideration is the noise the generator produces.

Standby generators generally operate in the range of 60 to 70 decibels. For frame of reference, normal conversation is about 60 to 65 decibels and a vacuum cleaner is in the 75 to 80 decibel range. Portable generators often operate at 90 to 95 decibels, the same as most lawn mowers.

Noise is an important factor for two reasons.

First, you may live in a community with noise ordinances. After all, leaf blowers have already been banned in many places—so who knows what's next.

Second, the noise from a generator, along with the smoke and smell, may act a magnet for anyone within earshot. It tells them that you have something they don't—electricity. They'll follow the noise as sure as a dog follows his nose to the frying pan, creating an operational security risk that you'll need to assess.

If you do use a portable generator, you can cut down on noise and fuel by not running it continually. Rather, cycle it on until the freezer reaches the desired temperature, then turn it off.

Before we move onto inverters, let me mention one other generator option for those of you with acreage.

If you have a 24 horsepower or larger tractor, a PTO generator is an excellent option.

For instance, there's a NorthStar generator that connects directly to your tractor's power take off (PTO) and provides 12,000 watts of power, more than enough to power your heavy duty appliances. Given that tractor fuel tanks often hold 25 gallons or more of fuel, diesel storage is easy. And, unlike generators with an engine, there's no engine maintenance or many parts to break on a PTO generator, since it uses the tractor's engine. Just be sure to have some heavy-duty extension cords that will reach from your tractor to the appliances you want to power.

Now, let's discuss inverters.

Method Two: Inverters—Whereas generators can be loud, inverters themselves are virtually silent, making them an excellent choice for apartments and maintaining operational security. The downside is that they have a very limited power supply, meaning you likely won't be powering a heat pump with them.

Many models are easily powerful enough to operate freezers, small microwaves and window air conditioners, however. For example, the Power Bright PW2300 Inverter cranks out 2,300 watts of continuous power and 4,600 watts of peak power. That's a lot of power, particularly in an emergency.

Inverters differ from generators in that an inverter only works if there is an *existing* source of electrical energy; it doesn't create its own. An inverter takes in direct-current (DC) power from deep cycle batteries or a car's 12-volt system—and converts it to the alternating-current (AC) power required by household devices, such as refrigerators and computers.

There are many situations where inverters make sense. For example, using a set-up of solar panels and deep cycle batteries with an inverter would be a good option for someone in a high-rise apartment, where using a generator or connecting to a vehicle battery would be difficult.

During Superstorm Sandy, Mario Aguilar Jr. used a Power Bright 6,000-watt inverter to run a refrigerator, lights and television in his New Jersey home. He simply connected the inverter to six marine batteries, which allowed him to get through the nine-day power outage with no food lost to spoilage.

Of course, using an inverter to power appliances will drain the battery that provides direct current to the inverter. You'll need a way to recharge the battery for as long as you plan to use the inverter.

The most common method of achieving that is simply connecting the inverter to a car or truck battery. By running the vehicle while the inverter is in operation, the vehicle's alternator will keep the battery charged and provide the electrical juice the inverter requires—that is, unless you put too much demand on the alternator.

Unfortunately, there are significant limitations with this approach, as each vehicle's alternator is sized to meet the demands of its specific vehicle. It is not designed to power juice-guzzling household appliances, such as microwaves.

So if you're planning on using an inverter connected to your car battery, opt for a small inverter that powers only a few low-energy devices.

If you need to use an inverter to deliver more power, some people connect deep cycle marine batteries or golf cart batteries to their inverters, and use solar panels to charge those batteries. You'll have to size the solar panels correctly for the system and, of course, find a place to locate the panels so they can capture the sunlight.

Whether you opt for a generator or an inverter, you first need to determine how much electricity you're trying to generate in order to properly size your inverter or generator, so let's look at the average wattage of a few common appliances.

- A coffee maker may draw from 600 to 1,200 watts.
- A small microwave draws about 1,000 watts.
- A refrigerator may draw 600 to 700 watts, as does a chest freezer.
- A space heater requires anywhere from 750 to 1,500 watts.
- A 42 inch TV may only require 250 watts or so, but you may want to power your satellite dish/receiver, which would require another 30 watts or so.
- Cell phone chargers, connected iPads or tablets, DVD players and clock radios each typically require 30 watts or less, some only using a few watts.

Then we get into the real power hogs.

- A 1/3 horsepower well pump may require 2,000 to 3,000 watts to start, and 750 watts to run continuously.
- A 1/3 horsepower sump pump requires about the same.

- A 7,000 to 10,000 BTU air conditioner needs up to 5,000 watts to start and 1,000 to 1,500 while running.

Of course these are all estimates. It's a simple matter for you to check the nameplate on each appliance you want to power so that you will know the actual wattage required. Alternatively, you can use a Kill-A-Watt.

Finally, you may not need to run all the appliances you think you do. In my case, I would prioritize running a freezer over the refrigerator and use alternate means for refrigeration.

One way of doing that is to simply store several well-insulated coolers in your preps, and to always keep several bottles of water frozen in two-liter bottles. When your refrigerator starts to warm after a power outage, put one or more frozen bottles into a cooler, and fill the cooler with items that require refrigeration.

By allocating power to your freezer and filling extra bottles with water (potable or not), you can continually add the frozen bottles to the cooler to create emergency refrigeration.

Whether an inverter or generator, or both, is right for you, take steps now to acquire them, learn to use them, store fuel or power for them and calculate how much electricity you need to produce.

Take a moment and figure out the best option for your needs.

CHECKPOINT—INVERTER OR GENERATOR

- Decide which option you prefer: inverter, generator, or both.
- Decide where you will procure one and where it will be stored.

- Decide if it will connect to your electrical panel via transfer switch or not.
- Write a summary of your current state, what you would like it to become and how you will get there.
- **Example**: I'll use my tractor's PTO generator for all my power needs. I keep the tractor full of 28 gallons of diesel, but will also store 50 gallons of diesel in 10 five-gallon containers, located in my workshop. I need to purchase three 50-foot heavy-duty extension cables to reach from the tractor to my appliances.

Ideally, you don't want to waste generator and inverter energy on tasks that you can accomplish by simply being prepared. So let's discuss ways to get water and light, and how to cook without electricity.

Water Without Electricity

Getting water when the grid is down will depend very much on your current water source.

In our case, we have a deep well. Well pumps require a lot of power, so we wired our pump to an outlet in the well house that can connect directly to a generator. If the power fails it's a simple matter to connect a small generator to the pump and keep the water flowing. This means we can continue to use our internal plumbing, making life a lot less stressful.

For long-term power outages when we may run out of fuel, we also have a FloJak Original deep well hand pump. If you too have a well, the FloJak is easy to install alongside your existing electric pump. It can pump pressurized water into your home for flushing, washing, and showering.

At about $500, these are not inexpensive. Then again, the average customer now pays an astonishing $123 per month for pay-TV, according to NPD Group. Since clean water is a little more valuable than reruns of Seinfeld in a grid-down situation, a FloJak or a similar pump may be a valuable investment. Just put your cable TV on hold for three months if you need the money.

So, our two primary backup sources of water are a generator to power our electric pump and a FloJak to manually pump clean drinking water to our house. If you have a well, you may want to mirror these approaches.

If, on the other hand, you have a municipal water supply chances are the water *may* keep flowing, though you won't have hot water unless you provide energy to your hot water heater.

If your municipal water source goes down you'll need to use methods described in chapter three to source your water.

Lighting Without Electricity

Since we've already discussed how you can produce electricity with inverters and generators, those are methods you can use to power lighting.

Often, using candles or oil lamps is the first thing that comes to mind for lighting without electricity. We have several Dietz oil lanterns that provide beautiful, nostalgic light. We stock lamp oil, but we can also burn cooking oils in the lamp, which is one reason why we keep large quantities of olive oil on hand. If the oils pass their expiration date, we simply store them as fuel for lanterns. If we're feeling fancy, we just infuse the oil with herbs or essential oils to add fragrance to the light.

Don't forget to keep plenty of wicks, matches, and lighters safely stored. Vacuum sealing boxes of matches is one method of ensuring they are dry when you need them.

If you'd rather stay away from fire, you can find both solar camping lanterns and hand-crank lanterns on Amazon and elsewhere. The Secur SP-1101 is an excellent hand-crank emergency spotlight and lantern that not only puts out light when and where you need it, but can charge smartphones and other digital devices through a USB port.

A clever solution for lighting during power outage is to use LED solar landscape lights. For instance, the InnoGear 2-in-1 LED landscape lights put out a whopping 200 lumens of light. Just keep the lights outside during the day to soak up the sun, and bring them inside each night to light your darkness.

Of course, you should have several good flashlights in your home and in your vehicles. For home, I recommend purchasing at least a few quality LED flashlights. When I say "quality," I'm not referring to the cheaply made LED flashlights often displayed at the checkout counter for less than $4 at convenience stores. We're talking about your disaster plan, here, and this is not the place to skimp. The low-cost internal parts of cheaply made flashlights can quickly corrode and fail just when you need them most.

A very popular budget choice for emergency kits is the AYL 3-in-1 LED Vehicle Emergency Flashlight. Priced at $20, it uses three AAA batteries and has multiple uses. In addition to being effective as a flashlight, the eight-inch device has a side panel lined with LED lights. The bright lights become a useful lantern, but clicking them again causes them to blink red, which is useful during roadside emergencies.

If you're interested in a better quality flashlight, I carry the Surefire G2X Tactical at all times, and can highly recommend it. You'll need to store batteries and chargers, of course, which is covered later in this chapter.

CHECKPOINT—GRID-DOWN LIGHTING

- Describe how you will produce lighting without electricity.
- Write a summary of your current state, what you would like it to become and how you will get there.
- **Example**: I will have six LED landscape lights outside that remain charged. If power goes out, we will need four in the house, one for each family member. I will also have two hurricane lamps and two gallons of lamp oil along with one hand-crank flashlight.

Cooking Without Electricity

I covered food storage in the prior chapter, and if the power goes out, you'll probably want a method for cooking.

For purists, nothing beats cooking over an open fire. Many preppers, us included, treasure Dutch ovens and cast iron cookware. You can buy them new, but you can also find them frequently at garage sales for a bargain. You may be surprised what you can use them for—we've cooked cakes and breads in the Dutch oven, and everything from steaks to fried chicken and cornbread and popcorn in the cast iron skillets.

Don't be put off by messy-looking cast iron pots and pans.

They can be cleaned and restored, just like new. In fact, there is an article to explain how to restore and care for cast iron cookware at selfsufficientman.com/castiron.

If you have a propane or charcoal grill, that method can serve you well in a power outage—as long as you have plenty of filled propane tanks on hand. If you prefer charcoal, consider filling a metal garbage can with charcoal and keeping it in a dry place, such as a garage or workshop.

If you don't have a charcoal grill, many preppers love the Volcano 3 Collapsible Cook Stove. It heats up very fast and is portable.

Likewise, the Kelly Kettle can create a blazing fire from just twigs and a spark, and comes with a pot to boil water or cook food.

We like to use camp stoves, so we have a Coleman InstaStart 2-Burner Stove. It uses the same small propane canisters that we use for the Mr. Heater portable heater. We store lots of propane canisters and use them for multiple purposes.

Some preppers are fans of solar ovens, such as the All American brand sun oven. These units simply use the power of the sun to reach temperatures up to 400°F, which is more than sufficient to bake, boil, or steam your favorite foods. But at $300, they're not cheap. If you need to save money and have the time to do this yourself, do a web search on "DIY solar ovens" and you'll find plenty of tutorials. Or, just visit solarcooking.org/ for plans and ideas.

If you live in a small apartment, some of these methods may not work for you, unless you have a patio. In that case, stock up on Sterno cooking fuel. These are the cans often seen under chafing dishes at buffets to keep food warm, but they can be used equally well to warm a can of food or cook a dish. Just have a pot with a lid to capture as much of the heat as possible. Each seven-ounce can burns for two hours, but can be extinguished if not needed for that long, then reused.

Regardless of what method you choose to cook, don't forget to store one of the most important kitchen tools: a manual can opener.

CHECKPOINT—COOKING WITHOUT ELECTRICITY

- Describe how you will cook without electricity.
- Write a summary of your current state, what you would like it to become and how you will get there.
- **Example**: I will use a Coleman camp stove to cook if we lose power. I will store 12 one-pound propane canisters in the garden shed. I'll use these to power both the camp stove and a portable heater.

BATTERY POWER

If the power grid went down for a very long time, such as months or years, I image that anyone without batteries would have an extremely difficult time adjusting to the "new" world. While I think it is wise for you to develop a plan for long-term power supply in a grid-down scenario, the focus of this book is to get you prepared for a short-duration crisis.

Achieving this is actually a simple matter by just storing enough of all the batteries you use to get you through a month. To calculate what you need, write down an inventory of the devices you use that require replaceable batteries, such as flashlights, cameras, and radios. Don't count remote controls unless you're planning on using them when the power is out.

If you're using non-rechargeable alkaline batteries, I recommend you switch now to using rechargeable batteries. We opted for an assortment of Panasonic Eneloop low-discharge nickel-metal hydride batteries (Ni-MH) and an Eneloop Super Power Pack charger. If our power fails for an extended time, we can easily recharge all the batteries we need by simply connecting the Eneloop charger to an inverter and battery set-up.

For electronic devices with non-removable batteries—such as cell phones, iPads, iPods, and laptop computers—a portable fold-up solar module is a great solution. Look for one in the 25 to 30 watt range. For example, the Instapark Mercury27 is a light and compact USB solar charging system. It combines four solar panels totaling 27 watts and folds up to about the size of a magazine, only thinner and lighter. This provides more than enough juice to power mobile devices and small appliances, and fits easily into a survival bag.

Note: 9-Volt Batteries can be a fire hazard!

> In 2012 the New Hampshire State Fire Marshal's Office issued a warning about how to locate and store batteries.
>
> In July of that year, a fire broke out in a kitchen "junk" drawer. The fire produced smoke throughout the first floor of the home. In the drawer were spare keys, a cigarette lighter, paper clips, eyeglass cleaner and some batteries in a plastic bag.

The fire department determined that a 9-volt battery ignited after rubbing against another battery stored in the same bag.

A 9-volt battery is a fire hazard because the positive and negative posts are right next to one another. If the ends come in contact with a metal object, such as aluminum foil, steel wool, paper clip, other batteries, etc. the object will heat up and can ignite a fire.

Store batteries in their original packaging or keep ends covered. For disposal, make sure that the positive and negative posts are safely wrapped in electrical tape—otherwise they could contact a metal surface in the trash and ignite.

CHECKPOINT—MAINTAINING BATTERIES

- List the essential devices you use that require batteries.
- think about how you will charge the batteries if the electrical grid is down.
- Write a summary of your current state, what you would like it to become and how you will get there.
- **Example**: I need AA and C batteries for flashlights. I will use rechargeable Eneloop batteries for this and will power the Super Pack Charger from my inverter, connected to my car battery. I will charge my IPhone and IPad with an Instapark USB solar charger.

Short-Term Fuel Considerations

If you have a vehicle or plan to use a generator or kerosene devices, you need a way to store fuel. How you do this depends on where you are located—a tiny apartment high above a city and a sprawling farm in the country provide quite different opportunities for fuel storage.

Whether you're storing diesel, gas or kerosene, make sure they are in approved containers, clearly labeled, and stored in a safe, dry, cool location. Most important, keep them away from any flammable source! We store our five-gallon cans in a shaded, detached workshop.

Keep each fuel container as full as possible to minimize the headspace, where condensation will form and add unwanted water to the fuel.

If you plan on storing fuel, understand that fresh, untreated gasoline may last only a year before it begins to deteriorate. For long-term storage that will last years, just add a fuel stabilizer such as PRI-G to gasoline and PRI-D to diesel.

Don't forget that one of the safest and best places to store gasoline is right in your vehicle's fuel tank. Try to *never let your car go below half a tank* before refilling. Yeah, I know, it's a pain to fill it more often, but it's one of the many things preppers do *just in case.*

Propane tanks and one-pound canisters should be stored upright in a well-ventilated area and not inside your home. Do not store propane in any part of your home or in a garage adjacent to your house. If a canister leaks while stored in a sealed room, such as a closet or basement, the gas can travel under doors and go undetected throughout the house. All that's necessary is for it to find a flame or spark—say in a kitchen or fireplace—to create a disaster.

Instead, store small propane tanks outside your house in a cool, well-ventilated location *below* the grade of the house. Keep it away from paths to lower areas, such as a root cellar or basement opening, since leaked propane is heavier than air and will descend to fill a hole. This could create a very dangerous scenario. Imagine someone tossing a lit cigarette into a depression filled with invisible propane.

Whether you're storing gas, propane or kerosene, your goal is to store enough fuel to get you through a short-term emergency. Calculate what you will need to get through two or four weeks, your choice, and figure out if you have the room to store it safely.

CHECKPOINT—FUEL STORAGE

- Decide what you want to provide fuel to.
- Determine what fuel type is required.
- Write a summary of your current state, what you would like it to become and how you will get there.
- **Example**: I need enough fuel to run my pick-up truck for two weeks so that it can power my inverter. My plan is to run the inverter for four two-hour intervals each day. This will require that I store ___ five-gallon cans in my workshop. I also need an extra 20 lb propane tank on hand for the gas grill.

FINAL THOUGHTS ON ENERGY

Many people will find it easier to address water and food storage than energy, but it doesn't have to be complicated. It can be as simple as having a few oil lamps and a fireplace with some cast iron pots—that's pretty much what grandma would have done.

I suspect you'll want more modern conveniences, however. For that, there's no escaping the fact that at least some investment is involved. Generators, inverters, and fuel aren't free, so if you want to enjoy the conveniences they provide, start saving right away. Put away a little money whenever you can toward the purchase.

Think of it this way: you'll likely lose more money through food spoilage in an extended power outage than with any reasonable inverter or generator investment. Given that power outages are pretty common, if you can find the money for an inverter or generator, it will not only give you peace of mind, it will likely save you money in the long term.

Also, why not go ahead and prepare yourself now for doing chores without power? Perhaps you could take a weekend twice a year or so and do everything manually—dishes, washing clothes, lighting and cooking without electricity, drying clothes, etc.

Since grandma taught me, I'll walk you through how to wash clothes without electricity in the next chapter.

When facing a grid-down disaster you will worry more about where to go to the bathroom than you will about what to eat.

Tim Young

CHAPTER SEVEN

PREPAREDNESS STEP 5: SANITATION/FIRST AID

When Superstorm Sandy pummeled New York, it provided a harsh reminder of how fragile our food, transportation or sanitation systems are.

Arriving at high tide, Sandy's strong winds created a record-shattering storm surge that overloaded the pipes and pumping stations in sewage treatment plants. Over 11 billion gallons of sewage then overflowed into rivers, harbors, canals, and some city streets. The sewage overflow was 50 times greater than the entire BP Deepwater Horizon oil spill in the Gulf of Mexico.

The storm caused massive interruption to a fragile infrastructure as it paralyzed transportation, flooded hospitals, crippled electrical substations, and shut down power and water to tens of millions of people.

In addition to 159 U.S. deaths attributed to the storm, a large number of families suffered cold, hunger, and extreme discomfort. No electricity meant more than no heat. It meant no running water, and no running water meant toilets stopped working.

Without electricity, sanitation became a nightmare overnight.

Let's face it—issues relating to sanitation and first aid do not make for pleasant dinner conversation, so it's understandable if you haven't given this much thought. Yet, for most of mankind's history, dealing with human waste, preventing disease, and treating illness was always an immediate concern.

So, taboo or not, we now have to discuss some messy subjects—human waste, cooking sanitation, and dirty laundry.

We'll start with human waste.

DEALING WITH HUMAN WASTE WITHOUT ELECTRICITY

Nature calls and—when you gotta go, you gotta go. But what if there is no water to flush toilets?

Other than digging the occasional cat hole on camping trips, pretty much none of us have ever handled our own body's waste. We just take a seat on a porcelain throne, unroll the toilet paper and flush the toilet. Poof—it's gone!

Nowadays the closest we get to dealing with our own waste is grabbing a plunger, which grosses us out.

As Sandy made all too clear, sanitation is a major concern in a disaster.

If you live in the suburbs or the country, your first inclination might be to "go in the bushes" when the water stops flowing. The problem is that your neighbors will think the same thing. This strategy risks starting an epidemic as, in some cases, more people die from cholera due to poor sanitation than from the initial disaster.

In a situation without running water, you'll have to manage human waste on your own. It's nothing to fear, as our ancestors certainly knew, but it does have to be handled *carefully*. No doubt many of us need a refresher—so pay attention.

If you have Municipal Sewer Service

With a municipal sewer system, your home has a pipe, called a lateral, that runs out to the street to connect with the sewer main.

You need to determine if the sewer main is DOWN or not. If you're not sure, check with your municipal water supply to confirm.

If the town sewer IS DOWN, DO NOT flush the toilet. The act of flushing will ultimately backflow sewage from the main into the rest of your plumbing system, including your tubs and sinks. If this occurs, you will likely have to evacuate your home.

Many homes have back-flow prevention valves to prevent this. You should find out if your home has one and, if not, consider installing one.

If the town sewer is NOT DOWN and the town's pumps ARE operational, follow the directions below for private septic systems.

If you have a private Septic System

Continue using your toilets as usual if you have running water.

If you don't have running water, fill the back tank of the toilet with water. This can come from a rain barrel, your swimming pool, or even a pond or creek. With "dirty" water, just filter it first through a T-Shirt or equivalent so you don't flush solids that could clog your system. Once you've put filtered water into the tank, flush as normal.

If these options do not work for you or are unavailable, you can use a portable sewage station.

PREPPING SUPPLIES FOR A PORTABLE SEWAGE STATION

The good news is that it's a relatively simple matter to prepare for disasters, even as it relates to sewage. We have it far easier than our ancestors did, as there are many sanitation and personal hygiene items available that you can add to your preps.

Some of these may not be completely necessary if you have a home and are able to follow the steps I outlined above, but they are convenient and offer additional back up in case you have to bug out. For everyone, they offer peace of mind.

Here are some items to consider for your sanitation station:

- Luggable Loo Portable Toilet: This is a very economically priced portable toilet. It's a bucket style with a traditional snap-on, hinged seat and cover, and is compatible with standard Double Doodie bags. That means there is virtually no cleanup and waste disposal is a snap.
- Simple DIY Toilet—For short-term emergencies, you can essentially make your own "luggable loo" for less than twenty bucks. You need nothing more than a five-gallon bucket lined with a heavy-duty garbage bag, a bag of peat moss and a toilet seat to cover the bucket. If you don't have a toilet seat, just place a couple of short 2 x 4s on top to serve as a seat. Place a thin layer of peat moss or cat litter in the bottom of the bucket. Then simply "go" in the bucket and add peat or cat litter as needed until services are restored. There should be no smell. Just don't let the bag get too heavy—you don't want it to break when you're disposing of it.

- Portable Emergency Zone Honey Bucket Toilet Set: This set includes liners and chemicals so you don't have to dig a latrine. A great choice whether you're staying in a fixed, temporary, or mobile location.
- Double Doodie Toilet Waste Bags: Double Doodie toilet waste bags provide easy, no-mess waste disposal. Each bag consists of an inner bag and an outer bag that is sealable and leak proof. This bag can be used with most portable toilets.
- Porta Potti 550E: This Porta Potti is the top of the line in portable toilets. It is sanitary, odorless, and leak proof.
- Composting Toilet: If you have the means, you may want to consider a Nature's Head Self-Contained Composting Toilet for both a disaster and everyday use.
- Toilet paper. Toilet paper may be the invention we most take for granted, and the one we may one day miss the most. Every prepper knows the importance of having lots of "TP" stored. In the aftermath of an economic catastrophe, it's possible for the production of toilet paper to run out. Stock up when things go on sale.
- Disposable Diapers. If you have a young'un you know all about these. Just make sure you have plenty on hand if you need them.
- Latex Gloves. Not only are they valuable in your first aid kit and in pandemic situations, but you'll need a barrier between you and contaminants in sewage disposal or other unsavory tasks.

- Hand Sanitizer. This kills germs but doesn't clean your hands—you'll want soap for that. Use hand sanitizer after using the bathroom and before handling food or beverages, so keep some disinfecting cleaning wipes and sprays on hand.
- Bleach. Sanitizing with bleach is effective, but bleach has a short shelf life. If you're not comfortable following the procedures for using calcium hypochlorite to make your own bleach, a good alternative is Steramine Quaternary Sanitizing Tablets. Just one small bottle makes 150 gallons of cleaning solution. Use it to spray on knives and butchering equipment, cutting boards, sinks, counter tops, and all other non-porous articles and surfaces.

DISPOSING OF HUMAN WASTE

On average, a person produces one ounce of stool per 12 pounds of body weight. In the case of a 192-pound adult, that works out to be one pound per day. Multiply that times two adults and two half sized children and a family of four would have to deal with three pounds of feces per day, or 42 pounds over two weeks.

Yikes!

If you have your own septic system, you can simply continue using the toilets by adding water to the toilet tank.

Without water for flushing in a sewer system, your toilet options are the portable sewage stations listed above or to use the cat hole method in the woods. The portable sewage station options work fine, but unless you opt for the composting toilet, you'll need to frequently dispose of your excrement.

So how do you do that?

To think of this topic more pleasantly, just think of it as camping in an RV.

First, separate urine from feces. This can be as simple as using a bucket with a lid to collect urine and a separate container for solids. Since urine is sterile, it can be diluted with 10 parts water and applied as a nitrogen-rich fertilizer directly to plants or lawns, if you'd like.

When using one of the portable sewage options such as a do-it-yourself toilet, the process is to defecate, scoop, and repeat. The "scoop" part can be kitty litter, peat moss, quicklime, or sawdust, and it's best if you add a little bleach solution each time.

After several uses it will be clear that it's time to dispose of the waste. The good news is that it's already bagged for you.

While urine is almost always sterile, feces contain dangerous pathogens and have to be disposed of properly. In a grid-down scenario, the solution is to dispose of the bagged contents by burial or burning.

If you live in an urban area, find out what the city wants you to do with the human waste. They will be trying to repair sewage facilities and should let you know where and how to dispose of the waste.

They might ask you to bring the waste to central dumping trenches or to bury it in your yard. For that reason, it is good to have some lime stored to layer over the waste in your portable sewage station, to help speed up composting. Store lime safely in childproof plastic buckets.

If you are unable to get advice from authorities or live in a very rural location, you may want to take your cue from the military.

A common military sanitation tactic is a burnout latrine. Simply place manure into an oil drum and burn it out daily by adding a little diesel fuel to it. Of course, this requires that you have a drum and stored diesel fuel, but by now you probably see how each of these 10 steps on the path to preparedness overlap and interconnect.

The other option is to simply bury your waste. This requires digging a simple trench, the size of which depends on how long it will be used and how many people it will serve. For a typical family, a reasonable start would be to dig 18 inches wide, 24 inches deep and several feet long. If you live in a rural area, a tractor or backhoe will come in handy during this time for you and neighbors. Otherwise, have some good shovels handy.

Locate your latrine at least 200 feet away, or about 70 adult paces, from your living quarters and water source. Be sure to choose a location at an elevation below your water source. For example, if you have a spring on your property, locate the latrine 200 feet away and below it, not above it.

Don't place the latrine in an area prone to rainwater runoff and try to locate it upwind from your prevailing wind direction.

When burying the waste, poke holes in the plastic bags with a pitchfork or other tool to aid in composting. Cover each layer with dirt and lime to deter insects.

Locating the latrine 200 feet away from your food prep area or water source will be very challenging in urban areas. If you opt to stay in those environments, you're left with the burnout option unless your community develops a better sanitation plan.

In a disaster situation, sanitation is critically important to prevent your family from getting seriously ill. Combined with a possible lack of medical facilities due to power outages, remaining healthy becomes even more crucial.

It's easy to see how sanitation can become an urban nightmare in a crisis that few people and towns are prepared to handle.

CHECKPOINT—HUMAN WASTE

- Determine if your choice of shelter relies on a septic system or municipal sewer.
- If sewer, determine if you have a back-flow prevention valve installed.
- Write down the phone number and address of where to find out if the sewer main is down.
- Decide now how you will dispose of human waste.
- Write a summary of your current state, what you would like it to become and how you will get there.
- **Example**: We have a private septic tank and will continue using our indoor toilets for feces. We will fill toilet tanks with filtered water from our swimming pool. We will not waste water to flush urine, but rather will collect urine in a five-gallon bucket with a lid. If we run out of water and there is a lengthy emergency, we will construct an outhouse 200 feet below our house and well.

DISEASES SPREAD THROUGH FECAL-ORAL TRANSMISSION

Since hands are the primary contact point with everyday objects as well as between people, they can spread disease if a person fails or is unable to wash after a bowel movement. Common houseflies can also spread disease by landing on feces and then checking out how well you cooked that burger you're eating.

Whether pests or people complete the fecal-oral route, ingesting even minute particles of fecal material can infect a person with any number of ailments, such as Rotavirus, Campylobacter, Hepatitis A, Salmonella, Cholera, Giardia, Typhoid Fever and more.

Given the many diseases that can be spread so easily, it's clear that sanitation is critically important during a crisis. As Chief Sanitation Officer, you need to prioritize sanitation and personal hygiene to prevent your family from suffering.

Now is the time to plan for how you'll handle that when the water and electricity stop flowing.

Earlier I described how to use calcium hypochlorite to make a stock of chlorine that you can use to disinfect drinking water. By following that approach you're actually making your own bleach.

To make a *hand sanitizer*, simply add one cup of the stock of chlorine to one gallon of water. Use this for cleaning surfaces and rinsing your hands after you have washed them.

PREPPING ITEMS FOR PERSONAL HYGIENE

For everything from brushing your teeth to taking showers, you'll want to store personal hygiene supplies. Of course, personal hygiene needs vary from family to family, ranging from sanitary napkins, diapers, contact lens care to even adult diapers. Get what your family needs.

The great thing about all these items is shelf life. Many of them will last for years, if not forever. Things that are perishable can be replaced with alternative items. For example,

- instead of storing body wash, store bar soap,
- don't store toothpaste—store baking soda, and
- store a bar of shampoo soap instead of bottled shampoo.

Within a year or so many body wash products will separate and darken in color. On the other hand, a bar of soap will last indefinitely if not exposed to air, so it's a better choice for prepping. The same holds true with toothpaste, which can take on an off taste and consistency over time. Baking soda, however, lasts forever and works great.

Remember to put aside extra toothbrushes and floss. If you now use electronic toothbrushes, be sure to purchase old-fashioned manual labor toothbrushes, just in case.

Regardless of whether you have children, be sure to store baby wipes. If the wipes dry out, simply place in a Ziploc bag with a little clean water and wait a bit. In a pinch, these are very effective for bathing, and require neither power nor water.

Speaking of bathing, enjoying a hot shower will be an extreme luxury in a disaster. One way you can achieve this is with an inexpensive Advanced Elements Summer/Solar Shower. Talk about a morale booster! A solar shower is a great way to keep clean and comfortable without running hot water.

PREPPING ITEMS FOR COOKING SANITATION

For cleaning in a disaster or grid-down scenario, you can reduce your reliance on water by stocking up on disposable plates and utensils.

Also, stock up on disinfectant sprays such as Lysol, which kills 99.9 percent of germs on hard, nonporous surfaces and is effective against over 50 microorganisms.

Stocking up on paper towels can provide relief from constantly washing towels. While they are marketed for automotive clean-up tasks, shop towels are great for wiping hands and cleaning up grease, oil, grime, and fluids.

CHECKPOINT—SANITATION

- Make a list of hygiene products that your family will require in the event you're unable to resupply at a store.
- Divide your list into two sections: perishable & non-perishable.
- Determine alternative products you can store to replace the perishable products.
- Write a summary of your current state, what you would like it to become and how you will get there.
- **Example**: My family will require the following perishable items: toothpaste, deodorant, and shampoo, but I will store baking soda and coconut oil to replace these. For non-perishable items, I will stock a supply of toothbrushes, bar soap, feminine napkins, disposable diapers, baby wipes, toilet paper,

shop towels, paper plates, plastic cups, plastic utensils, Ziploc bags, heavy duty trash bags, aluminum foil, and dish soap

PREPPING ITEMS FOR LAUNDRY

If you don't like doing laundry with modern conveniences, you may really hate doing it without water or electricity! Still, it's an important task that awaits your attention, and it will be easier if you plan for the event now.

Why is fresh laundry so important? Because dirty clothes can harbor microorganisms, and wearing clothes with these microorganisms can lead to skin infections.

Dirty underwear has traces of germs from body fluids and defecation, which makes infection more likely if it's worn again without being laundered. The best way to prevent the spread of germs found on clothes from normal wear is to wash them.

My grandma had no trouble washing clothes without electricity, so you can do it too. Here's what you'll need:

- A washboard or some rough surfaces to get out the main dirt.
- One or more galvanized metal buckets.
- Bars of laundry soap such as Fels Naptha.
- A clothes wringer for clothes. Alternatively, loop clothes around a clean railing, pull tight and twist the water out.

- A clothesline and clothespins, or a clothes rack. Stock up on these items at any "dollar" store. There are even indoor clotheslines for apartment dwellers.

HOW TO WASH CLOTHES BY HAND WITH A WASHBOARD

- Prepare your laundry by separating into lights and darks.
- Try to locate a free-flowing and clear water supply from a stream or pond if you can. Otherwise use water from a cistern, water tank, or bottled water. Stagnant water, such as swimming pool water when the pump isn't working, will likely harbor bacteria, mosquito larvae, etc., which defeats the purpose—unless you run it through a Big Berkey or similar water filter.
- Washing with hot water isn't always required, but if you need to, heat water in a galvanized tub over a hot fire. Or, if it's a hot summer day, fill the tub with water in the morning and let the sun heat it during the day.
- Add detergent to the water if you have it. Otherwise, grate laundry soap or even a regular bar of soap. If you have nothing else, a tablespoon or two of dishwashing soap can also do the trick.
- Add your laundry, whites first, and agitate thoroughly with your hands, a wooden paddle or a laundry plunger such as the Rapid Washer. Do this for several minutes, depending on how soiled the clothes are.
- For set stains, lay the item on a washboard set over the tub and rub soap on the stain.
- Rub the item back and forth over the board until the

dirt or stain is removed.

- Rinse the clothes two or three times in clean water until there's no more soap left. This is why it is desirable to have at least two tubs.
- Wring out with a Best Hands clothes wringer or by hand, then hang to dry.

Since washing clothes by hand is a lost art, let me offer a couple of tips.

To prevent stiffness, beat the clothes with a thick rod or stick occasionally while they dry. Beating the clothes may also help relieve stress you feel because the power outage is causing you to miss The Simpsons reruns.

You can also give the clothes a firm shake before hanging to help reduce the stiffness.

Another tip is to make your own laundry softener. Just mix one part vinegar, one part baking soda, and two parts water. Then, add to the rinse water.

CHECKPOINT—LAUNDRY

- Visualize yourself doing laundry by hand.
- Go through each step and make a list of items you will need.
- Write a summary of your current state, what you would like it to become and how you will get there.
- **Example**: I can use our plastic kiddie pool to hold warm water that we will heat on a grill. I would like to purchase a washboard, but in the meantime I can gather a rough rock from our property that, once

cleaned, offers a surface for scrubbing. I will need another large container for rinse water but a simple Rubbermaid tub will work for that. For now, I can use it for storing laundry supplies such as extra detergent. I'll also store baking soda and vinegar to use as softener. Fortunately, I already have a clothesline and clothespins.

FIRST AID

When times are good, as I hope they are for you while you're reading this, it's easy to get pills and bandages from the pharmacy or to get to an emergency room if needed. But what if times change, even for only a couple of weeks, and you or your young child need medical attention?

What then?

During a catastrophe the first responders (police officers, firefighters or paramedics) may not be available, while doctors or nurses will likely be overwhelmed. In that case, *you are your family's first responder.*

Many prepping and preparedness books delve into advanced medical and first aid topics. Often I see videos of preppers-in-training practicing surgery on a dead pig's foot with an army surgical kit, expecting that they may have to perform an emergency appendectomy when all hell breaks lose.

For the record, I've never operated on a pig's foot, dead or alive.

Since the aim of this book is to help you *start prepping*, let's focus on the common medical needs you may have to attend to.

These could include cuts, burns, broken bones, allergic reactions, venomous bites, choking, risk of drowning, poisoning and so on.

Just as we're used to flushing away our waste, we've become accustomed to calling 9-1-1 for any emergency, so the thought of having to treat a child or loved one ourselves is a frightening scenario to ponder. But in a disaster, you may not only be the Chief Sanitation Officer, you will be the family's Chief Medical Officer.

Whether you have any medical training or not, we all have to accept this responsibility. The good news is that there is a lot of knowledge available and more ways to access it today than ever before. It's simply a matter of committing yourself to study, but do it now while there is no emergency.

Here are four steps you can take to increase your medical preparedness.

Step One. Get the Right Books. The topic of survival medicine can be a book on its own, and often is. I recommend having the following books on hand *in print*:

First Aid Book

- The Special Operations Forces Medical Handbook is an excellent resource. It includes nearly 140 comprehensive illustrations that show the proper techniques for medical care, from basic first aid and orthopedics to instructions for emergency war surgery and even veterinary medicine. You can even download a free pdf version at tinyurl.com/SOFbook.

Medical Books

Here are three excellent books for when there is no doctor:

- The Survival Medicine Handbook: A Guide for When Help is Not on the Way, by Joseph Alton. Now in its second edition, this thorough guide assumes that no doctors or hospitals are available due to a short-term or long-term disaster scenario, and that YOU are now responsible for your family's health.
- When There Is No Doctor: Preventive and Emergency Healthcare in Challenging Times, by Gerard S. Doyle. This well-designed book is full of medical tips and emergency suggestions.

Step Two. Get hands-on training from the American Red Cross. Take an American Red Cross first aid class or CPR class. Learn to recognize and care for a variety of first aid emergencies, such as burns, cuts and scrapes, sudden illnesses, head, neck and back injuries, and heat and cold emergencies. Often the classes take only a couple of hours, so take a course and then go home and teach your family.

Don't forget to check with your local fire and police departments. Some offer first responder classes.

Step Three. Be prepared to know how to use everything in your first aid kit. Allergic reactions, asthma attacks, poisonings, and snakebites require immediate action. You'll only know how to handle these urgent first aid situations if you prepare before they happen. *This is serious stuff so read the books I just referenced.* Only by doing so can you become fully prepared for the medical situations you may have to face on your own. Pay particular attention to treatments for

- anaphylactic shock,
- asthma or chronic lung conditions,
- bites from venomous snakes, spiders, and scorpions,
- shock,
- burns,
- choking,
- drowning,
- blisters,
- bleeding,
- tourniquets and broken bones,
- using ammonia inhalants, and
- cuts and wound care.

Step Four. If a picture is worth a thousand words, a video may be worth a million. Study free virtual training such as:

- The Patriot Nurse on YouTube. The Patriot Nurse provides excellent tips for first aid and medical concerns as well as practical prepping advice. Visit youtube.com/thepatriotnurse
- Doom and Bloom Survival Medical Channel. From the authors of *The Survival Medicine Handbook*, learn from hundreds of videos on survival medicine and prepping at youtube.com/drbonespodcast.
- The Survival Doctor Website. Author and survivalist James Hubbard, M. D., M.P.H. offers guidelines for life-threatening emergencies at survivaldoctor.com.

- The Mayo Clinic First Aid Index: The Mayo Clinic has an extremely useful First Aid Index to review and print for your personal survival manual. Go to mayo-clinic.org/first-aid

All this new knowledge will be overwhelming, of course, so it's best to try and absorb it in small doses. Make a habit of watching some videos weekly—perhaps devoting a couple of hours at the same time each week. If you include all family members in watching videos, then you can learn together.

With some basic knowledge under your belt, you can then pack a medical kit.

SURVIVAL MEDICAL KIT SUGGESTED CONTENTS

Without professional emergency assistance, a well-stocked medical kit can mean the difference between life and death in a survival situation. A bad one not only offers inadequate resources, it can be inherently dangerous for someone to use if they're not adequately prepared. Therefore, study the survival medical resources listed above until you understand when, why, and how to use your medical supplies.

You'll probably need more than one medical kit. We keep a fully stocked kit in our home and another in the car. Most of the supplies for the kit are very inexpensive, and you may already have many things you need in your medicine cabinet. Just organize them and know how to use them all—if you don't, read a tutorial or watch a video online. If you don't have Internet access, visit a library that offers it.

The Red Cross recommends that first aid kits for a family of four include

- 2 absorbent compress dressings (5 x 9 inches),
- 25 adhesive bandages (assorted sizes),
- 1 adhesive cloth tape (10 yards x 1 inch),
- 5 antibiotic ointment packets (approximately 1 gram),
- 5 antiseptic wipe packets,
- 2 packets of aspirin (81 mg each—can be administered under the tongue immediately in the event of a heart attack),
- 1 blanket (space blanket),
- 1 breathing barrier (with one-way valve),
- 1 instant cold compress,
- 2 pairs of non-latex gloves (size: large),
- 2 hydrocortisone ointment packets (approximately 1 gram each),
- Scissors,
- 1 roller bandage (3 inches wide),
- 1 roller bandage (4 inches wide),
- 5 sterile gauze pads (3 x 3 inches),
- 5 sterile gauze pads (4 x 4 inches),
- Oral thermometer (non-mercury/non glass),
- 2 triangular bandages,

- Tweezers, and
- First aid/medical booklet.

In addition to the Red Cross recommendations, I suggest adding the following.

- Pick a good container—I recommend a tackle box, as the slots are just the right size for bandages, wipes, etc.
- Snake bite/sting kit.
- Anti-diarrheal medicine.
- Blister packs.
- Eye patches.
- Fish antibiotics (make sure they're made in the U.S. and make sure you know when/how to administer).
- Finger splints.
- WaterJel burn dressing.
- Ammonia inhalant.
- Bar of soap or antibacterial lotion.
- QuickClot Clotting Sponge.
- Israeli First Aid Compression Bandage.
- Stretch tourniquet.
- Steri-Strip skin closures.
- If you or someone in your family is on medication, try to store an emergency supply. If you can't get your doctor or pharmacy to issue you an extra supply, try

refilling each prescription three or four days early each month. After several months you'll amass a 10 to 14 day supply of medicine for an emergency.

- An old cell phone that has enough battery power to turn on. Cell phones don't need a service contract to call 9-1-1, though they do need a working signal. So take an old cell phone you don't use anymore and keep it in your first aid kit for emergencies.

CHECKPOINT—MEDICAL

- Identify the books, websites, and courses you want to rely on for learning first aid skills.
- Go to your medicine cabinet and assess your supplies.
- Determine what else you need to make a complete medical kit.
- Get duplicates of necessary items and create a second kit for your vehicle.
- Write a summary of your current state, what you would like it to become and how you will get there.
- **Example**: I purchased a pre-made first aid kit for my car, but it lacks a space blanket, breathing barrier, snake bite kit, ammonia inhalant, QuickClot and Israeli Compression Bandage, so I will add those. For my home I will add everything recommended in this chapter to a tackle box I have. I'm not familiar with using fish antibiotics, so I will research that.

FINAL THOUGHTS ON SANITATION & MEDICAL

I'm afraid the area of sanitation and medical is one that most people give little thought to until a disaster strikes. In our modern society, we rarely have to think about such things, since clean water always flows toward us and our wastewater always flows away from us—to parts unknown.

I know this is hard to think about; not just having to deal with your own body's waste, but the fact that your normalcy bias convinces you that this will never be a serious problem. Sure, it may happen, but only for a day or two.

Or so you think.

When facing a grid-down disaster you will worry more about where to go to the bathroom than you will about what to eat. After all, while you may not prefer to, you can go weeks without eating and even days without water.

Try holding your bowels that long—heck, many of us can't even finish a car trip without pulling over three times.

So give the issue of sanitation some serious thought, PLEASE.

Fortunately, most people have given at least a little thought to medical preparedness, but that often goes no further than tossing an emergency kit into the trunk or closet.

Ask yourself—"What would I do if my son came in with a gushing laceration on his abdomen right now and there was no way to get emergency help?"

Think about these issues now while times are good so you can be prepared when they aren't.

Don't forget your pets

Depending on the pet, you may need a litter box, paper towels, plastic trash bags, grooming items, and/or household bleach.

Keep their medicines and medical records stored in a waterproof container. Have a first aid kit handy with a pet first aid book or smart phone app, as I suggest later in the book.

If violent crime is to be curbed, it is only the intended victim who can do it. The felon does not fear the police, judge or jury. Therefore he must be taught to fear his victim.

Jeff Cooper

CHAPTER EIGHT

PREPAREDNESS STEP 6: PERSONAL SECURITY & THE SURVIVAL MINDSET

This is the longest chapter of the book and may be the one that pushes you the most out of your comfort zone. In terms of the 10-Step Path to Preparedness, you may find it the most difficult to put into practice, because there are few lists to check off or specialty items to buy. Rather, you will be challenged to change the way you think and break some lifelong habits.

None of us like to think about the "what if" scenarios in which our family members are harmed—or worse. For many, it's too hard to accept that human nature can really be capable of violence, so we escort these thoughts from our head and deny that violence can happen to us.

Normalcy bias tells you that violence happens to someone else.

I need you to stop forcing those thoughts out of your head. Instead, I want you to embrace those "what if" thoughts for what they are—not visions of things to come, but rather signs that you take preparedness seriously. A sign that you're aware the world is full of risks—that there are some violent people, some hazardous situations, some horrific accidents—and you don't want to be a victim.

The good news is that if you implement the suggestions in this book that make sense for you, you'll find yourself fully prepared for short-term emergencies and widespread disruptions.

The bad news is that will make you an extreme minority, and the envy of many hungry, unprepared souls.

Disasters do often bring out the best in people, as neighbors help one another to cope with tragedy. For the *most* part people tend to pull together and help one another in times of natural disaster.

Unfortunately, disasters also bring out the worst in people, particularly those who already have a "shady" mindset, or who become desperate. Desperate people do desperate things (imagine what you would do to feed your children) and if you want to keep your family safe, you must be prepared to protect them as well as your preps.

How will you defend yourself when unprepared thugs want the food in your stocked pantry? Or when violence breaks out as you and others wait in a long line at the town's last working gas station?

And it's not just disasters we have to be prepared for. Violence is all around us, happening all the time.

We're shocked when we see headlines of violence, such as

- Bombing at the Boston Marathon
- Gunman kills 26, Including 20 Schoolchildren in Connecticut
- 13 Dead, 29 Injured at Fort Hood Shooting
- Gunman shoots 70, Kills 12 at Batman Movie
- Virginia Tech Shooting Leaves 33 Dead
- 9 Killed at Church in Charleston, SC

We keep ourselves glued to the news reports and cringe when we contemplate the truly horrific nature of the tragedy. But do we change our behavior? Do we do anything to prepare ourselves for such an event? Do we take our children aside and teach them skills that could save their lives?

For most people, the answer is no. We turn our heads and remain inactive.

John Leach, author of *Survival Psychology*, sums up the problem with this behavior when he says:

"Denial and inactivity prepare people well for the roles of victim and corpse."

So our collective strategy seems to be to hope for the best as a relentless wave of violence washes over us. Even in "good" and "stable" times, look no further than these typical 2015 headlines for examples of urban powder kegs ready to explode:

- 28 people shot in Baltimore over Memorial Day weekend; city breaks homicide record.
- 56 shot, 12 killed in violent Memorial Day weekend in Chicago.

Two cities. Eighty-four people shot over the course of a weekend. And this is in *good* times.

In the past, one could expect to be sheltered from civil unrest by living in a rural community rather than a large city. This is no longer the case, for our rural youth are suffering even more than their urban counterparts. According to the medical journal *JAMA Pediatrics*, rates of death from suicide among children, teens, and young adults are nearly double in rural communities compared to urban areas.

Though they may be burdened by a lack of opportunities and limited social support services, today's rural youth are more connected to the world than ever. They know of protests, uprisings, and violence instantly via tweets, Facebook updates, and YouTube videos. Given the right circumstances, these updates can fuel the spark for every corner of the country to ignite simultaneously, and without warning.

You may not think violence or civil unrest can come your way, but it can—and faster than you can react unless you have a survival mindset.

From home invasions to muggings to road rage and church shootings, violence surrounds us. But it's also not just violent criminals we have to prepare for—it's everyday events, where life can change in a split second.

LIFE OR DEATH IN A SPLIT SECOND

In 2002 and 2003, I spent a lot of weekends playing guitar in a classic rock band throughout southern New England. Nothing serious, but we did have a decent following of fans who enjoyed hearing southern rock classics made popular by Lynyrd Skynyrd, Tom Petty and the like.

One of my jobs was to book gigs for the band, and we were excited when I booked us a gig at a prominent rock club in West Warwick Rhode Island. The club was called The Station.

A few weeks before our scheduled date, the 1980s rock band Great White played The Station on February 20, 2003. Even though the club's legal capacity was only 404, over 500 people attended. I wasn't among them, but Bob Young, a former employee of mine, was. He stood six feet, six inches tall, and everyone just called him Big Bob.

By the time Great White took the stage just after 11:00 p.m. the nightclub was jam packed with fans longing to relive the '80s.

Just seconds into the band's opening song, the band's manager set off pyrotechnics, as planned. What wasn't planned was the flammable soundproofing foam behind the stage being ignited by the sparks.

Sadly, videos taken by fans indicate that most people thought the flames were part of the act, so they muddled about. Many kept talking, holding their drinks and pointing to the flames.

The fire spread remarkably fast as it raced along the walls and ceiling, producing intense heat and a thick, toxic, black smoke as melting foam dripped from the walls. The panicked crowd finally raced toward the narrow exits, which quickly became clogged with bodies lying on top of one another.

Within a couple of minutes the entire building was fully engulfed in flames.

Bob Young died in the fire, along with 99 others. Most of the victims died at the main entrance where the frantic rush created a logjam at the front door. The 100 deaths made The Station fire the fourth-deadliest nightclub fire in U.S. history. Hundreds more were injured.

It's awful to watch, but you can see a graphic, unedited video of The Station nightclub tragedy at tinyurl.com/thestationfire. Viewing it could help you understand how you may be able to survive tragedies simply by increasing your situational awareness and altering the way you assess danger.

As you increase your situational awareness, you'll begin to see the world differently by spotting risks in places you never saw them before. But how do we change the way we see the world without our social circles banning us because they view us as a ranting gloom and doom buzz kill?

We become aware. We prepare. And we practice the survival mindset.

THE SURVIVAL MINDSET

The best form of personal preservation isn't a weapon. It's a mindset.

Some people are fortunate to be born with an instinctive survival mindset, while others have to work hard to overcome their lack of situational awareness. But it CAN be learned. It must be learned if you hope to be prepared to protect your family and what's yours.

Let's start with what not to do.

Many people are blind to situational awareness. Perhaps you've seen them walking with their heads down or only looking straight ahead, never behind, to their sides, or up or down. You've likely noticed an increasing number of adults and children walking while texting or looking at their iDevices, oblivious to what's happening around them.

They are easy targets for crime and violent attacks.

You're not one of those people—are you?

The late Colonel Jeff Cooper developed a Color Code system for situational awareness. This system is not a physical process, but a mental one that should be used whether a person is armed or not. It teaches how to be alert and avoid a deadly confrontation, which should always be the goal. After all, as Sun Tzu said:

"Ultimate victory is in avoiding the fight."

Cooper's Color Code system is based on four colors, with white representing the least threatening state and red representing the most imminent danger. Between them, the threats escalate first to yellow and then to orange.

I believe understanding how to use this tool to master situational awareness is far more important than what type of gun, ammo or self-defense weapon you have.

Bottom line—this system can save your life.

Let's take a look at how this works:

WHITE	YELLOW	ORANGE	RED
Relaxed	Relaxed Awareness	Specific Alert	Mental Trigger
Unaware	Aware of Environment	Anticipation	"If he, then I..."
Unprepared	Easy to Maintain	Full Attention	Fight or Flight

Condition White: You are unaware and unprepared. If attacked in Condition White, the only thing that may save you is the inadequacy or ineptitude of your attacker. When confronted by something dangerous, your reaction will probably be "Oh my God! This can't be happening to me."

Condition Yellow: Relaxed alert. No specific threat situation. You are simply aware that the world is a potentially unfriendly place and that you may have to defend yourself, if necessary. You use your eyes and ears. You don't have to be armed in this state, but if you are armed you should be in Condition Yellow. You should always be in Condition Yellow whenever you are in unfamiliar surroundings or among people you don't know. You can remain in Condition Yellow for long periods, as long as you are able to watch your rear. In Condition Yellow, you are taking in surrounding information in a relaxed but alert manner.

Condition Orange: There is a specific alert. Something is not quite right and has your attention. Your radar has picked up a specific alert. You shift your primary focus to determine if there is a threat while continuing to watch your rear. Your mindset shifts to "I may have to defend myself against that person," focusing on the specific target that has caused the escalation in alert status. Cooper suggests that in Condition Orange, you set a mental trigger: "If that person does 'X,' I will need to stop them." If you have one, your pistol usually remains holstered in this state. If the threat proves to be nothing, you shift back to Condition Yellow.

Condition Red: Condition Red is fight. Your mental trigger established in Condition Orange has been tripped. "If 'X' happens I will fight that person" — "X" has happened, so you must now engage.

What the Color Code system really comes down to is this—pay attention to what's going on around you at all times. Then deal with the situation you observe as it REALLY is and not as you wish it to be or fear it might be. You MUST overcome your normalcy bias.

Let me describe how I follow this system.

I may be in a Condition White state while inside my own home or that of a family member or close friend, as long as I know and completely trust everyone there. I can't think of too many other situations where I would be that relaxed and completely at ease.

I would remain in Condition Yellow if I were in a public place, such as ball game or restaurant, assuming that I was seated in a corner (my preferred seat) with clear access to a convenient exit. Generally speaking, I practice at Condition Yellow being the lowest level of my awareness. Anytime I carry a firearm I aim for Condition Yellow. If I'm at a sporting event or concert, I watch the field or stage to enjoy the moment, but constantly scan the crowd around and behind me for anything threatening or suspicious.

I escalate to Condition Orange if I noticed something out of the ordinary—perhaps someone eyeing me from across a street or looking over my vehicle as I walked out of the convenience store. Perhaps a fight breaking out a few rows behind me at a ball game. There are many things I could notice that would cause me to escalate from Condition Yellow to Condition Orange, such as

- an unusual or unnatural obstacle on the road in front of my vehicle or blocking my walking path,
- a vehicle following me,
- finding myself suddenly in a dense crowd with limited escape routes,
- another person's pace matching mine as I walk in a public place or city street,
- someone taking an angle that intercepts my path,
- suspiciously concealed hands,

- a heavy or long jacket when weather doesn't warrant it,
- darkness and limited visibility when I'm exposed,
- absence of people and a "feeling" of vulnerability,
- hearing people argue nearby, or
- seeing someone intoxicated in public—especially if there are two or more people.

Condition Red is bad news, and long before I reach that point my senses are very heightened. Condition Red means that I've identified a clear threat that I have to deal with immediately. I'm facing a fight or flight scenario, but most likely fight. I would likely have chosen flight prior to reaching this state unless I'm a victim of a sudden attack.

I don't want to be in Condition Red, but if I find myself there I'll be thankful for my everyday carry items, such as a bright flashlight, pepper spray, and a loaded firearm, even though I'll do everything in my power to evade and not be forced to draw a lethal weapon.

Reaching Condition Red is *not* the time to wonder what the laws are in your state regarding self-defense, so *find that out in advance*. For example, many states have stand-your-ground laws. According to Wikipedia, stand-your-ground laws

> "Authorize a person to protect and defend one's own life and limb against threat or perceived threat. This law states that an individual has *no duty to retreat* from any place he/she has a lawful right to be and may use *any level of force*, including lethal, if he/she reasonably believes he/

> she faces an imminent and immediate threat of serious bodily harm or death; this is as opposed to duty to retreat laws."

I'm sure you'll agree that the goal is to never be in a situation that requires you to defend yourself or your family. If you're careful about your surroundings and escalate through the stages of awareness, chances are good that you'll never encounter a Condition Red situation.

However, if you walk out of the mall carrying an armful of shopping bags with your head down, you're what a criminal may view as low-hanging fruit, so don't put yourself in that situation, and train *all* family members to be diligent.

Naturally, it's expected that you'll oscillate between these mindset states.

For instance, let's say that I'm in a city at a sidewalk cafe seated outside. I'm reading the paper with my headphones connected, to eliminate the street noise. While I'm actually reading I may momentarily be in Condition White, but I make it a point to not look down to read, but rather to hold the item up so that my peripheral vision can detect anything out of the ordinary. Even if I spot nothing unusual, I make it a point to go Condition Yellow and scan the area every couple of minutes.

IMPROVING YOUR SURVIVAL MINDSET

My wife and I have a routine of relaxing in front of a movie most nights—it's our favorite way to unwind after a day of homesteading. If the doomsday prognosticators are right and Armageddon is on the horizon, movie time may be what I miss the most from our modern lifestyle.

One of our favorite movie characters is Jason Bourne, brought to life by Matt Damon in the "Bourne" movies. In *The Bourne Identity*, the first movie of the series, there's a scene in which Damon's character, Jason Bourne, demonstrates a remarkable survival mindset. At this juncture of the movie Bourne is struggling to recall his identity, having suffered amnesia two weeks prior as a result of a traumatic experience. After entering a restaurant he sits at a table with Marie, a woman he just met and to whom he is paying $20,000 for a ride to Paris. Here is part of the dialog between the two:

> Jason Bourne: I come in here, and the first thing I'm doing is I'm catching the sight lines and looking for an exit.
>
> Marie: I see the exit sign, too. I'm not worried. I mean, you were shot. People do all kinds of weird and amazing stuff when they are scared.

Bourne pauses for a moment as he tries to figure out who he is and why he thinks the way he does. Then he continues:

> "I can tell you the license plate numbers of all six cars outside. I can tell you that our waitress is left-handed and the guy sitting up at the counter weighs two hundred fifteen pounds and knows how to handle himself. I know the best place to look for a gun is the cab of the gray truck outside, and at this altitude, I can run flat out for a half mile before my hands start shaking. How can I know that and not know who I am?"

Bourne may not know but we do. He wasn't born with such a remarkable survival mindset. Rather, he trained himself to have remarkable situational awareness.

And you can do the same.

Okay, sure, that's is a Hollywood production and it may not be reasonable (or necessary) to remember the license plate numbers of every car in the parking lot. But it is necessary to be alert and pay attention to your surroundings. It's especially important to quickly assess the threats (people or other potential threats) when you enter a room and to know how you can immediately exit that room if you need to. It could also save your life.

But you're not Jason Bourne. You're an average Joe, or Betty. So how do you go about improving your survival mindset?

Start by listening to your gut.

Trust Your Feelings

I believe that we all have "feelings" and "gut instincts" that serve us well, but many of us may have lost touch with them, or simply don't trust ourselves. Your survival instincts are deeply ingrained and you're smarter than you think you are. Don't overlook your instincts just because of their intangible nature.

In a potential survival situation, listen to and rely on your instincts. If you pay attention you'll likely know or "feel" when you're in a potentially dangerous situation.

At The Station nightclub, alarming clues before the concert began could have been the few and very narrow exits and the difficulty reaching them due to overcrowding.

When walking into any building that will confine you, pay close attention to the location of designated exits, such as doors and stairwells, as well as exits you could improvise, such as fire escapes, or low-level windows you could shatter.

If you find yourself in a very crowded situation, pay close attention to where you are positioned relative to the exits and have a plan to escape if you need to. This includes restaurants, hotels, office buildings, or any other place that is potentially "trapping" you.

Position yourself near the safest location at all times. The view from the top of hotels is often quite nice, but I always request a room on the lowest floor possible for an obvious reason—it's the fastest way out. When I get to my room, the first thing I do is look for the exits—not just one, but multiple. If a fire, act of terrorism or anything else occurs while I'm in a strange place, I want to know my options.

Once I had to quickly evacuate a fire threat at a Glasgow hotel, but was able to get out of the building in seconds since I had requested a first-floor room.

Having a finely tuned survival mindset may be your most important preparedness skill, and the good news is that, unlike most preps, it doesn't cost you anything. It's simply a skill that you and your family should practice every day. So practice it and teach your children to do the same. It could save your life, and theirs.

Here are some tips on how to do that:

- When visiting public or new places, always identify the emergency exit locations and make sure everyone in your group does the same.
- Look at the size of the room and the number of occupants. Ask yourself if everyone could get out at the same time if there was a sudden bomb scare, fire, gunman, or similar emergency.
- When sitting in restaurants and other public places, try to choose a seat with your back to a solid wall,

such as a corner. This will allow you to face others and provides maximum visibility.

- Practice walking with your head high and scanning your horizons as you walk. Pay careful attention to not fall into a predictable pattern of simply looking left to right—what if the threat is above or below? Rather, learn a relaxed ability to simply take in everything in your range of vision, remembering occasionally to look over each shoulder. You'll soon become highly proficient at identifying risks.
- Now that your head is held high, focus on your body language and attire. Do you walk confidently? Do you look alert? Are you dressed in shoes that give you flexibility to move quickly if necessary, or are you wearing flip-flops? If you're carrying items, such as a purse or shopping bag, are they held close and secure? *Don't look like an easy target for anyone*!
- While driving, continually think of the best place to steer the car off the road should you have an immediate threat. For instance, when I'm driving I pay running attention to whether there are embankments, trees, guardrails, or flat areas to my left or right. Should an oncoming vehicle inexplicably and suddenly turn in my direction, my hope is that I'll react a fraction sooner by having this knowledge.
- Pay close attention to where you park your car. Make a point to park in highly visible spots, such as under bright lighting or where there is a lot of human traffic.
- Carry a flashlight with you at all times. When you come out from the movie late at night, you'll be glad

you did as you light your path, scan your surroundings and, if needed, shine a blinding light into the eyes of a potential threat.

- Don't be robotic. What do I mean by this? Let's say that you're following your GPS in a strange area. You're in a city on a busy street, but the GPS steers you down a narrow, dark alley, even though your destination isn't close. Should you robotically follow your GPS? No. Rather, stop, look, and think about what you're doing. The alley may be perfectly safe, but the point is to think for yourself and not let someone else (or an electronic device) think for you.
- Speaking of electronic devices—PUT THEM DOWN! Encourage everyone in your family to pay attention to their surroundings and not to their digital screens when they're walking, whether to your car or on a street.

Whether you want to think of it as the survival mindset or situational awareness, I believe this is the most critical preparedness skill you can possess. By practicing this every day, you can and will keep yourself out of dangerous situations. Learn how to be aware and teach your children the same! And remember: don't fight with your fists. Fight with your brain.

CHECKPOINT—SURVIVAL MINDSET

- Take a good hard look at your beliefs and habits.
- Evaluate your current survival mindset and level of situational awareness.

- How often are you in Condition White vs the other color levels?
- What is your level of awareness when you are in public? How do you present yourself? What do your clothes and behavior say about you and your ability to react? What do your mannerisms say? Where is your attention focused?
- How often do you think through potential "what ifs?"
- Write a summary of your current state, what you would like it to become and how you will get there.
- **Example**: Right now I block out visions of "what if" because they frighten me. I often wear clothes unsuitable for getting out of trouble, such as high heels that make it difficult to escape. I leave my purse in the shopping cart as I go to unlock the car door, and I even text on my phone sometimes while walking through parking lots. I will immediately change these behaviors, starting with how I dress. I'll assume that there may be a need to "make a run for it" so I'll dress appropriately. I'll try to always be aware of what color level I'm at and will make a point to go to Condition Yellow frequently.

Now that you have an understanding of the survival mindset, let's examine personal security.

PERSONAL SECURITY

I don't want to scare you, but you may very likely have to protect yourself and your family in a disaster. After any crisis some individuals take advantage of the chaos by robbing and harming others. They know that emergency resources are swamped with the disaster at hand, thus lawlessness and disorder ensue.

When a disaster strikes, police and other emergency personnel will be overwhelmed. Some will simply walk off the job to tend to their own families, as over 200 New Orleans police officers did during Hurricane Katrina.

Just as most individuals lend neighbors a helping hand, some will simply choose to help themselves. There will be some looting, stealing, and an overall increase in acts of violence. There always is.

I mean, for goodness' sake, just look at the headlines about what happens in stores on the day after Thanksgiving, also known as Black Friday.

- Stabbing - Two Men Arrested In Stabbing Over Walmart Parking Space.
- Dying Man Ignored—A 61-year-old man collapsed and died while shopping at a Target store in West Virginia. Fellow shoppers ignored him, even stepping over his body while they continued to hunt for bargains.
- Toys "R" Us Terror—When two woman got into a bloody brawl at a Toys "R" Us in Palm Desert, Calif. on Black Friday, their husbands pulled out guns and began shooting in the aisles, and killed each other.
- Pepper Spray Panic—On Black Friday 2011, a shopper at a Los Angeles Walmart began spraying the crowd with pepper spray to get an advantage in getting a discount Xbox.

- Marine Mauling—A U.S. Marine collecting Toys for Tots donations at a Best Buy in Augusta, Georgia was stabbed on Black Friday 2010.

Pushing, shoving, fistfights, and even shooting just to save a few dollars on plastic toys. If so many people can easily descend to violence due to road rage or discount shopping, imagine how quickly the thin veneer of civilization will be stripped off in a real crisis.

Wherever you live, now is a good time to find out more about your neighbors and their abilities so you can incorporate that into your planning. Even a well-armed person will have a hard time defending against a violent gang intent upon taking his or her possessions. In that instance, odds for survival will increase substantially by aligning with neighbors for mutual defense—see chapter 12 on creating prepper teams.

Personal security isn't simply one thing, such as owning a weapon. It has several components, so I've broken this chapter into covering them. We've already addressed situational awareness and the survival mindset, which I believe is the most important aspect of staying safe in good times and in bad. In the remainder of this chapter I'll address child safety, operational security, home security, and family protection.

Before we get to those, let's discuss how you can protect yourself from the very real threat of violence.

Personal Protection

Let's begin by addressing a topic that is frequently debated, often with fervor.

Let's discuss firearms.

I'm not going to advise you on whether to own a gun or not. Actually, I'm going to recommend you NOT own a firearm unless you're committed to getting expert training and practicing with it frequently. Using a firearm effectively is a perishable skill; don't think you can be trained once and then be done with it. Otherwise you run the risk of doing more harm to yourself than if you didn't have a gun.

The choice to own firearms is a personal one, and you'll need to carefully consider what is right for you and your family. As you mull it over, however, take a moment to ponder the following scenario:

- **You're asleep in your bed when you hear glass shatter in the living room, two rooms away.**
- Your spouse bolts straight up beside you as you both hear heavy footfalls along with multiple loud male voices, evidently in an attempt to incite fear.
- Your six-year-old daughter and three-year-old son are in their bedrooms upstairs.
- You have three seconds to react to this situation.
- **What will you do to save yourself and your family from harm or death?**

How would you respond? Would you call 9-1-1, which could take a half hour or more for help to arrive? Would you simply freeze? Would you grab a baseball bat or something else lying around?

For me, and for many prepared people, the preference is to have loaded firearms accessible at all times, and to be highly proficient in using them.

Conversely, many people are uncomfortable having firearms in the house, particularly when children are present. Citing statistics on accidental shootings, Gavin de Becker, author of the excellent book, *The Gift of Fear*, says,

> "it's easy to conclude that, for most families, having a gun in the house increases risk."

Yet researchers at the Cato Institute reviewed eight years of news reports about shooting in self-defense and concluded,

> "the vast majority of gun owners are ethical and competent, and tens of thousands of crimes are prevented each year by ordinary citizens with guns."

In the Cato report, "Tough Targets: When Criminals Face Armed Resistance from Citizens," the authors argue that there are likely thousands of incidents when the display of a legal weapon *prevents* a crime. Since news stories only report when an armed citizen actually shoots a criminal, these "non-crimes" don't make the news or impact the statistics.

Regardless of your view on owning firearms, it's undeniable that if you have one or more in your possession and are expert in using it, you stand a greater chance of being able to protect yourself and what's yours. The chief virtue of firearms is their stopping power and ability to project force over a great distance, without you having to engage hand to hand. So in that regard, it may be worthwhile for you to consider firearms.

Tips for Those New to or Uncomfortable With Firearms

- NRA courses are often geared for very beginners and some are for women only. The courses were very helpful in teaching my wife to become comfortable with firearms.

- Ask a friend who hunts or shoots competitively to give you some firearm basics.
- Find a buddy with whom to take a basic firearms course.
- Watch DVDs on firearm safety in the privacy of your own home.
- Learn the laws in your state concerning firearms and self-defense, as well as the laws in any state you may carry a firearm to.

Self-Defense Alternatives to Firearms

If you're not comfortable with firearms or are unable to legally carry them, you're left with alternatives that put you closer to your aggressor.

Before listing these alternatives and the pros and cons of each, I need to point out that it is *very important* for you to know three things about each of these options.

1. Your state's laws regarding owning and using them.
2. How to use them effectively.
3. How to keep them safely protected from children or anyone you don't want accessing them.

The weapon you choose should also meet four criteria.

1. It must be easy to carry.
2. It must be simple to deploy.

3. It must be effective.
4. It must be legal.

It will do you no good to have a weapon that you don't know how to use and that an assailant can easily remove and use against you. Just as firearms require a serious commitment to training, so too do most of these alternatives to firearms.

Pepper Spray—Most people are familiar with pepper spray, but a common misconception about it is that it is primarily a weapon for women. In reality, most police officers, male and female, carry pepper spray on their duty belts.

And for good reason.

It doesn't have the range of a firearm, but pepper spray can give you a little distance, generally spraying up to 10 feet. I like the Sabre 3-in-1 Advanced Police Strength, which is a compact size.

This Sabre model can fire off 25 bursts, meaning you can do a quick test burst twice a year to test its operation. Even when I carry a concealed firearm I carry pepper spray since it is an effective non-lethal means of personal protection. It works equally well on non-humans, such as dogs and bears.

If pepper spray is your self-defense weapon of choice, don't just buy a canister and tuck it in your pocket or purse. You need to know how to use it. Is the spray pattern a stream, mist, or a cone? How far is it effective? How long should you fire it? When you get your pepper spray, test it so you'll know exactly how to use it when it counts.

While the effect of pepper spray is normally instantaneous, it may take a few seconds to overcome an aggressor. When training with your pepper spray, practice "spraying and moving," where you sidestep from your spraying position. That way if the person blindly attempts to overpower you, you've moved out of the way.

The downside of pepper spray is that, unlike using a firearm or striking someone over the head with a blunt object, it does not prevent motivated attackers from remaining a threat. Your goal after spraying is to evade and seek help, if possible. If it is not possible, you must prepare to physically defend yourself. You'll probably know that before spraying, so spray as much as you can and kick hard in vulnerable spots until you can reach safety.

Alternatively, simply plan on having pepper spray in your weak hand and deploying a more blunt weapon in your strong hand, such as a baton.

Collapsible Batons—Another excellent self-defense choice is to carry a Collapsible self-defense baton. I carry one anytime I'm unable to carry a concealed firearm, such as when I cross into a state that doesn't have reciprocity laws recognizing my concealed permit.

Whereas an attacker may be able to fight through pepper spray simply by holding their hat or another object to block the stream, a well-placed baton strike could fracture their hands or render them unconscious. A baton extends your reach so that you can strike first without fear of being hit, but batons can be taken away. Don't think this will be an effective weapon for you if you haven't practiced with it.

A baton can appear intimidating. Just imagine someone with an extended metal baton walking aggressively toward you. That's the way you'll want to appear if you're forced to draw it—confident and not an easy target.

Stun Guns and Tasers—These two weapons are often mentioned in the same breath, but there is a big difference between them.

Stun guns must be applied directly to the target. You hold the stun gun in your hand and, while pushing a button, touch the electrical nodes directly to a person. Stun guns make a loud, threatening sound, but to use them you must be in direct contact with the person.

By contrast, a taser fires two small electrodes up to 15 feet. A taser is more powerful than a stun gun and will immobilize the victim as long as you keep the current flowing. The downside is that you only get one shot. If you miss when firing the taser, you're out of luck. Also, it won't help you deal with multiple attackers.

Knives—I carry a folding knife, as many people do, and even a four-inch blade can be a formidable weapon in the right hands. In the wrong hands, it's easy to take away and use against the owner. Again, as with any option, it's not about carrying a weapon—it's about having the skill to use it.

High Intensity Light—A defensive weapon often overlooked is a high-intensity flashlight, which is one of many reasons I always have a powerful flashlight on me. Fighting off someone who is blinded by a bright light to the eyes is much easier. If you have an opportunity to flee, it's much easier to do so after flashing a light into an attacker's eyes.

Zap makes a 1 million volt combination flashlight/stun gun that combines two self-defense weapons into one. It's bulkier than I like for everyday carry, but it's an excellent tool for my wife to carry in her hand while walking to her car at night.

Canes & Walking Sticks—By simply making a habit of walking with a solid cane or walking stick, you become superbly armed. You can even seek local martial arts training to teach you how to turn these sticks into deadly weapons. Or if you'd like to practice this in your home, there is a DVD called Stick Self Defense offered on Amazon.

Keep in mind that you're not always trying to defend yourself against people. Often the attacker is a menacing dog. Canes, stun guns, and sprays work well against these threats.

Of course, you always have the option of blunt objects such as baseball bats or whatever is lying around. If you have excellent situational awareness you will have already surveyed the room to find weapons you can improvise—telephone cords, hot beverages, ordinary writing pens, etc.

If you have no other options you may be left with your open hands or combat training such as Krav Maga or mixed martial arts (MMA). Don't discount them—these can be used very effectively to eliminate a threat. However, they put you in close hand-to-hand confrontation with the aggressor.

Hand-to-hand is too close for my comfort and certainly not where I want my wife and daughter to be, so we maintain alertness and carry weapons to defend ourselves.

The choice is yours, but if you can't protect what you've put aside, why bother?

Refuse to be a victim. Be prepared.

CHECKPOINT—PERSONAL PROTECTION

- Evaluate your level of personal protection ability. Determine where you need to improve and if you need skills or tools to help you get there, such as MMA or a weapon.
- Think about various situations where you could face threats and determine how you would like to respond to them, if forced.

- Write a summary of your current state, what you would like it to become and how you will get there.
- **Example**: I'm comfortable with firearms in my home, so I keep a loaded 12-gauge shotgun where I can easily get to it. I've spent a lot of time training my daughter how to shoot, so I'm comfortable that she knows the power and purpose of this weapon. Still, I keep it out of her normal reach. I'm not comfortable carrying a handgun, so I carry pepper spray at all times. I wish I had physical self-defense skills, so I'm going to look for a Krav Maga class close by. If I find one I'll take my daughter with me and we'll do it together.

Rate your risks

Are you interested in seeing what your personal risk is of violent crime? Murder, burglary, or serious assault? Ken Pence of Vanderbilt University created an illuminating website that will tell you. As the rateyourrisk.org website says:

> "Crime exists. Your vulnerability cannot be ignored. Threat assessment is a means for you to calmly evaluate your risks. The online tests will let you realistically determine your chances. Close the door when you take the rape, robbery, stabbed, shot, beaten test so you give yourself a fair assessment. These tests give you a 'ballpark estimate' on your risk and are meant to entertain while educating."

TEACHING CHILDREN ABOUT PERSONAL SAFETY

Children need to know they are safe, and we parents want more than anything to ensure they are. One of the most troubling aspects of our current society is the horrific acts of violence we hear about in the news, often targeted at our children. These unspeakable tragedies include abductions, sexual assaults, and deadly shootings at schools, from kindergartens to universities.

One of the most awful realizations for a parent, at least for this parent, is that we cannot always be there to protect our children. Therefore, the best we can do is to calmly but consistently teach them the skills they need to stay safe, and I firmly believe this topic has a place regarding prepping and preparedness.

An enduring myth regarding safety for children is that the threats to children come from strangers. In the majority of cases, the perpetrator is someone familiar to the parent or child, and that person may be in a position of trust or responsibility to the child and family.

Adults should seek opportunities for "teachable" moments to introduce and reinforce personal safety skills to children. If an incident occurs in your community or on the news, and your child asks about it, speak frankly but with reassurance. Explain to your children that you want to discuss personal safety with them, so they will know what to do if they are ever confronted with a difficult situation.

One way I do this with my young daughter is to discuss this while she watches a favorite animated movie, such as Finding Nemo. There are a few potentially dangerous scenes in that movie, such as when a predator fish attacks and eats all eggs other than Nemo in the opening scene.

My young daughter has an adorable habit of putting her hands over her ears while keeping her eyes open anytime she witnesses something that alarms her, and this visual cue tells me when I have an opportunity to reassure, but educate her at the same time. As appropriate, I explain that any creature can be a predator, whether it's a fish, chicken, fox, or human. We're never too young to learn the difference between good guys and bad guys. Look for situations that are right for you and your children.

Here are some ideas that I hope will be helpful:

- Speak to your children about personal safety in small doses at first. Use a calm, non-threatening manner so you don't scare them but rather inform them that, while most people may be good and trustworthy, not everyone is. As they become more comfortable, create drills to increase their awareness of their own instincts, when to say no, when to run and seek help. If you don't think you can do this, you can. You're their parent and it's your job.

- Speak openly about safety issues. Children will be less likely to come to you if the issue is enshrouded in secrecy.

- Don't confuse children with the concept of "strangers." Young children don't have the same understanding as an adult might of who a stranger is. The "stranger-danger" message is *not effective*, as danger to children is much greater from someone you or they know than from a "stranger." Of all children under age five murdered from 1976 to 2005—31 percent were killed by fathers, 29 percent were killed by mothers, 23 percent were killed by male acquaintances, seven percent were killed by other relatives. Only three percent were killed by strangers.

- Speaking of children, know where they are—at all times. If you can't be with them at all times there are many real-time GPS trackers available for children so you can monitor their location and receive alerts if they go out of approved areas. Newer devices, such as one by hereO, simply act as a colorful watch that young children love.
- Older children and teens may think they already know all this or they are "too cool" to be lectured. They aren't, and teens are equally at risk from victimization.
- Children need to know that they can safely tell you or a trusted adult if they feel scared, uncomfortable, or confused at all.
- Above all, teach your children that it is more important to get out of a threatening situation than it is to be polite.
- An example of a video that may help you to inform young children is at tinyurl.com/iceabduct. In this social experiment video created with parent's permission, an ice cream truck operator demonstrates how frighteningly easy it is to abduct children. As a father of a young daughter, it's hard to watch, but what could be more important than teaching our children about safety? The time to begin developing their survival mindset is as soon as possible.
- Practice what you talk about. You may think your children understand your message, but until they can

incorporate it into their daily lives, they may not clearly understand it. Find opportunities to practice "what if" scenarios, such as reinforcing the survival mindset color drills mentioned earlier. For younger children, be sure to make it playful and fun.

- Get the family working out together and valuing fitness and skills as a lifelong habit. This could be martial arts, for example.

I believe it is critical that we discuss the issue of safety with our children, and do so frequently. Not to scare them, but to heighten their awareness of the world that surrounds them.

To demonstrate that introducing these survival concepts to children doesn't have to be scary, let me show you how to have some fun.

PREPAREDNESS THROUGH PLAYTIME

Unlike most adults, children stay in a constant state of learning just by being kids. By nature, skill building is in their bones.

From a biological, evolutionary perspective, the primary purpose of play is to promote skill learning. Play is nature's way of assuring that young mammals, including young humans, will practice and become good at the skills they need to develop to survive and thrive in their environments.

In his excellent book *Free to Learn*, author and developmental psychologist Peter Gray says that,

> "Play is activity for its own sake, not activity aimed at some serious goal such as food, money, gold stars or praise. When we offer such rewards

> to children who are playing, we turn their play into something that is no longer play. Because play is activity done for its own sake rather than for some conscious end, people often see play as frivolous, or trivial. But here is the deliciously paradoxical point: Play's educational power lies in its triviality."

In other words, play is nature's way of teaching us the skills we need for life. As parents, we must make conscious choices for what those skills are. It's our role to orchestrate the balancing act of allowing children to play with ensuring the skills they practice are the survival skills we want them to become proficient in.

There's an appropriate message for children regardless of their age, and an appropriate way to deliver the message. You're the parent—you know best, but by improving your own survival mindset you will model and teach them theirs just as surely as you taught them how to look both ways before crossing a street.

For example, you could teach a preschooler the skills to plan a survival bag simply by making up a game. Tell them to pretend they can't take the bus (or car) home from school (or playgroup) and you get to walk. Let them help pack the backpack. Many toddlers will no doubt contribute life-sustaining plastic dinosaurs and goldfish crackers, but this gives you an opportunity to show them a map, discuss sun, shade, and rain, as well as food and water. On your pretend journey, which could be from the pretend school in your garden shed back to the child's bedroom, let them decide if the sun is blazing hot or if you're trapped in a sudden downpour. They make the rules of play; you simply play with them and practice an important skill, such as taking shelter, making a play tent and so on.

There are countless games you can play with young and older children, and I've included a list below to get you going. As they develop their survival skills and broaden their thinking, you can share with your children why you bought the extra cans of tuna and why you're showing them how to filter water. They're young—they'll get it. Preparedness play should be fun, never scary.

As your children grow, you can build on the skills they've learned so they have a fighting chance in the event they have to face an emergency on their own. This, of course, is your end goal—to prepare your children for life.

Below are eight games you can use to teach situational awareness skills to children. I'm sure you'll have fun and can think of more of your own.

Game: Concentration.

Lesson: Situational awareness.

Approach: Have your kids look at an area, a person, a vehicle or an object. Then ask them to close their eyes and tell you everything they can remember.

Game: Guts

Lesson: Develop an awareness of "gut feelings."

Approach: Start by asking your children how they feel about certain situations—what is their gut telling them? Do they feel safe? Are they worried? Are they scared? Find places to show them things that you think look suspicious, and tell them why. Whether it's at an amusement park ride or in a crowded place, you'll find moments to explore their feelings.

For example, "I feel slightly nervous because the roller coaster goes so high and fast—but I feel excited too." Encourage them to dig deep—ask them why they feel that way. Ask them to look at the faces of people coming off the roller coaster so the child can weigh the risk of the ride against the reward. Most important, teach your children to trust their OWN instincts and avoid anything and anyone that seems suspicious.

Game: I-Spy

Lesson: Teach situational awareness.

Approach: Feel free to incorporate standard games such as "I Spy." For example, I spy an exit when you go into a restaurant. I spy a water source. Look for non-threatening ways to teach the survival mindset color chart.

For instance, I mentioned earlier how the opening scene in Finding Nemo is perfect to illustrate a quick transition from white to yellow to orange and then, tragically, to red. There are many teaching opportunities like this in your normal daily life—look for them.

Game: Hide and Seek.

Lesson: Find safe spots to hide.

Approach: All children know and love this game, but play with them so you can help them learn great places to hide, how to stay still and silent and when it is okay to come out.

Game: Bug Out!

Lesson: Practice evacuations.

Approach: You'll be able to come up with lots of ways to do this in an age-appropriate manner. For my young daughter, I might simulate a bedroom snowstorm in a small area where her dolly lives. I'd use white crinkle paper or something similar, but the point is that dolly needs to bug out—and fast! Is she prepared? Does dolly have her bag of dresses and bottles packed and ready? Where will she go? As she gets older we may pretend we have three minutes to pack up everything we'll need to evacuate. This will teach her to be ready at the drop of a hat in case something happens. Then, we'll bug out to the local ice cream stand for an emergency treat.

Game: What If?

Lesson: Contingency planning.

Approach: A great game to play while driving or anywhere. I'm sure you'll think of lots of questions depending on the age of your child. Just keep it non-threatening and be sure they're having fun. For example, what should we do if a blizzard is coming our way? What should we do if we arrive home and the front door is open? If a window is shattered? If the smoke alarm goes off while we're sleeping—and so on.

Game: Freeze!

Lesson: Make better choices.

Approach: Let's face it: sometimes your children are going to make poor choices as it relates to personal safety. Rather than scolding them, simply say, "FREEZE" when you spot behavior you'd like to discuss. For example, your young child may begin to walk ahead of you in a parking lot, or your teenager may walk heads-down while sending a message. Teach them to stop anytime you say, "FREEZE" and then ask them what's wrong with what they're doing. If they can figure it out and correct their behavior, perhaps it's worthy of a reward.

Game: Bring Your A Game.

Lesson: Teaching situational awareness.

Approach: The "A" stands for awareness and this is a game that you, as a parent, play without the child really knowing. Play by challenging your children at every opportunity to assess their awareness. For instance, while standing in line in the bank ask, "What's the name of this bank? How did we get here? How many ways are there to exit?" After a party or play date you may simply ask, "Who was at your friend's party? How many adults were there? How many were men and how many were women? Who did you not know?" If you have teenagers, you can empower them to play this game with their younger siblings, thereby teaching both at the same time.

In addition to these eight games, I describe 18 others in the eBook, *Playful Preparedness*. You can download it for free at selfsufficientman.com/playful-preparedness, or you can purchase it on Amazon.

If your child is addicted to technology, why not direct that passion toward learning preparedness skills? Ready.gov has online games that help children build survival bags, and play the part of the hero in earthquakes, fires and natural disasters. You can find the games online at ready.gov/kids/games.

American Red Cross has an application that you may want your kids to play. Monster Guard is a mobile app designed specifically for kids aged 7 to 11. Children can follow Maya, Chad, Olivia and all the monsters and learn how to prepare for real-life emergencies at home and other environments in a fun and engaging game. This is a free app available to download on iOS and Android devices. You can learn more at redcross.org/monsterguard

---

CHECKPOINT—CHILD SAFETY

- Observe your child and determine what safety skills you would like them to learn.
- Use one of the games in this book or develop your own game to help them learn.
- Be sure to make it fun—that's the essence of playtime.
- Write a summary of your current state, what you would like it to become and how you will get there.
- **Example**: My son has a wonderful, outgoing personality, but he is too trusting. I want to introduce him to the concept that some people and situations are dangerous, but he

is only four years old. I'll start by playing The A Game with him any chance I can.

Operational Security (OPSEC)

I find OPSEC to be a tough one. I understand why so many preppers preach OPSEC and to not tell anyone that they embrace preparedness. After all, that leads to people saying things such as, "*When the crap hits the fan I'll just come to your place.*" And those people seem to mean it!

So preppers attempt to maintain OPSEC and say nothing. I suspect the phrase "loose lips sink ships" has endured in our culture for a reason.

The trouble is that it's hard to really do that, particularly if you have young children, who are free to tell others what mommy and daddy are doing. After all, it is a good idea to involve children in the prepping process, particularly in the area of familiarizing them with how to use certain devices, evacuation and communication procedures and so on. However, it's important to discuss with your children the need for your family's privacy.

In the "old days" OPSEC wasn't a big deal, as it was well known that most people had a stocked pantry. Those who didn't would find help from others in the community who did. I can assure you that grandma never heard the term OPSEC.

Today, I figure that less than one percent of households have enough food to last a couple of weeks, making it unsafe for you to announce that you have a deep larder. So my advice is

- don't announce that you have emergency supplies,
- keep your supplies hidden so that visitors don't ask

questions, and

- quietly introduce neighbors and family members to preparedness principles by sharing this book and other resources to see what they think and if "Maybe we should prepare for the unexpected." The more people around you who are prepared, the less likely they will be to bang on your door—or break in.

In the prepper world there is a concept known as "The Gray Man," which simply describes an unassuming person dressed in drab gray who blends perfectly with the crowd on the street. Your goal is to not drive a yellow Humvee, to not show people your preps, but to rather simply be The Gray Man in the community.

Why might a prepper purposely wear dirty laundry?

a prepper may find it desirable to camouflage in society by appearing as dirty as the others are, so as not to attract unwanted attention. That's when wearing the dirty clothes may become an important prepping strategy. talk about becoming the gray man!

Home Protection

There are a number of reasons why you may be the only one who can protect your home, possessions, and family members, whether there be a disaster or not. After all, even in normal times it takes time for emergency help to arrive. The more prepared you are to take responsibility for your life and possessions, the less likely you are to be a victim of tragedy.

You'll need to think this one through for yourself, as everyone's home and environment is unique. Here are a few things to think about to get you started with home protection:

- Be sure to have working multi-purpose fire extinguishers clearly visible throughout your home, but *especially in the kitchen*, where most house fires originate.
- In some scenarios you may not have telephone access and the fire department will be swamped with other priorities, so you'll need the tools and the SKILLS to suppress fires on your own. If you live on a rural property, you may want to have the ability to pump water from a property pond or, better yet, have excellent water pressure from a cistern supplied by a gravity-driven spring to do so.
- If you have time to plant them, thorny bushes or vines below windows are helpful for deterring intruders.
- If you have a multi-story home, remember to have roll-out ladders to escape fire danger.
- Take a walk outside your home and look at the points of entry as if you were locked out and needed entry. What windows or doors are easy to get to? If they're easy for you, they're easy for others.
- A simple way to make your doors more secure is to replace the standard strike plate screws, which are typically one inch or less, with three-inch screws. This will make it extremely difficult for an intruder to pry loose the strike plate.
- Install perimeter fencing around your home. Having a strand or more of barbed wire on top makes it less inviting to potential intruders.
- Once dusk arrives, be sure to close curtains and lower

blinds throughout the house. Otherwise, it will be easy to observe the brightly lit interior from outside.

- Buy several American Red Cross Eton Blackout Buddy Emergency LED Flashlight/Nightlights and keep plugged into your electrical outlets. If the power goes out you'll have emergency lighting and won't have to fumble around looking for flashlights.
- Speaking of dusk and darkness, solar landscape lighting can be very useful for two reasons: First, it can light outside areas at night, making it less attractive for potential criminals to carry out an intrusion. Second, in a grid-down scenario, whether short or long term, these lights can be brought inside each night to provide interior lighting. LED lights, such as the Moonray Richmond Solar Lights, are much brighter than traditional solar lights, lightweight, and easy to carry around the house without the risk of accidental fire.
- Of course you can have a monitored alarm system, but even in normal times you'll have to wait for help to arrive. A better bet is to have loud audible alarms to scare away intruders, such as the Doberman SE-0104 Infrared Home Defender. This infrared motion activated detector triggers a loud 100-decibel alarm if an intruder trespasses into the protected zone. The device sells for about $20.

Regarding my own property security, we have a very long (half mile) private driveway, fenced property and gate, driveway access over a narrow bridge and solar motion lights. We also depend on several loud, big and aggressive guard dogs to deter uninvited visitors and announce their presence long before they arrive. Dogs, unlike some security systems, work just fine when the power is out.

Guard Dogs

There is one critical prepping item for family protection that I would not want to be without, and that is having at least one family guard dog.

I realize that guard dogs may not be an option for everyone. Some people live where dogs are not allowed. Others may have allergies. However, if you *can* have a dog and don't have one, I recommend that you consider getting one or more. Guard dogs can help protect both your possessions and your family. There are many breeds that are dependable and known for possessing the characteristics needed to ward off intruders.

The characteristics I look for in a guard-dog breed are

- friendly with a good temperament,
- loyal,
- an excellent watch dog,
- intimidating to intruders, and
- good with children.

There are many excellent breeds that expert dog trainers frequently recommend. These breeds include Bullmastiff, Doberman Pinschers, Rottweilers, German Shepherds, Komondors, Rhodesian Ridgebacks and others.

I'm no dog whisperer, but I'd like to suggest two other breeds that I have a lot of experience with: the Anatolian Shepherd and the Great Pyrenees.

We have crossed these breeds for years and use several as *live-stock* guardian dogs. They are also wonderful family dogs, but bark and charge fiercely towards anything attempting to penetrate the property that doesn't belong. That includes hawks, owls, coyotes, people on foot, and people in cars.

Having said that, my personal favorite for a *family* protection dog is the German Shepherd, the breed most commonly used by police forces. While German Shepherds can sit quietly in a home at night until a threat requires their attention, the Anatolian Shepherd and the Great Pyrenees often will bark the night away to deter any threat from even considering approaching the property. So for a house family protection dog, I lean towards the German Shepherd.

Regardless of what breed is right for you, a family protection dog will give you years of companionship and protection, and provide an alarm system that will give you time to react to any home threat.

If for some reason you can't or don't want to have dogs, you can still *appear* to have their protection. Just put up a sign at the entrance of your property warning about your dog, and then place a large dog bowl on your front porch. It won't help you if someone breaks in, but perhaps it can deter some potential intruders from making the attempt.

CHECKPOINT—HOME PROTECTION

- Review your home's security by walking around the outside with the intent of breaking in.
- Identify opportunities to make your home appear more secure, and be more secure.

- Devise a plan of what each person in your family will do if there is an emergency in your home. Practice a drill.
- Write a summary of your current state, what you would like it to become and how you will get there.
- **Example**: Our home has many low, easy-access windows. We have curtains on all of them to pull at night and have deadbolts on each door. we have motion-sensor lights outside, but nothing to stop someone from walking right up to a window. Thorny bushes would be good. Our family plan for emergency at home consists of my husband evaluating the threat (ex. where smoke is coming from, intruders, etc.) while I grab my cell phone and go to our son's room. My son knows to wait for me there with the door closed.

FINAL THOUGHTS ON THE SURVIVAL MINDSET

In closing this chapter I'd like to tell you a story about two pigs. Well—two breeds of pigs.

I've raised lots of pigs and one of the breeds I've raised is called Large Blacks. These friendly hogs grow very large and have big, round floppy ears. The ears actually drape softly over their eyes, which makes it difficult for them to see. So the Large Blacks move very slowly and even 500-pound hogs are easy to handle. You can walk right up on a pair without them noticing.

By contrast, we also raise a breed called Ossabaw Island hogs, one generation removed from wild pigs. In any group of two or more Ossabaws the pigs will stand alert, noses in the air and eyes facing opposite directions—watching each other's back or "six," as we say. No matter how sneaky I try to be, they skedaddle long before I get close to them.

It's next to impossible to sneak up on an Ossabaw—or any wild pig for that matter. They have a great survival mindset and stay in Condition Yellow or Orange at all times.

You don't want to see them in Condition Red. I've been on livestock trailers with a group of males when they feel threatened. They attack fiercely with their tusks.

On the other hand, the Large Black breed of pigs, like too many people, languishes in the oblivion of Condition White. I can put a leash on their neck and walk them onto a trailer and drive them to the slaughterhouse.

The moral of the story?

Be the Ossabaw.

"The trick is to stop thinking of it as 'your' money."

IRS Auditor

CHAPTER NINE

PREPAREDNESS STEP 7: MONEY & DOCUMENTS

While I strongly advocate abstaining from debt, this is not a chapter about avoiding debt, budgeting or anything like that. It is about having reserves so you will be financially prepared for emergencies.

Financial *reserves* mean just that; having *reserves* safely kept *in your possession* that you can easily access, without relying on a bank.

To be clear, I'm not suggesting that you should not keep money in banks or investment accounts. What I'm recommending is that you keep an adequate supply of cash on hand, ideally three months living expenses but a minimum of one month. After all, the aim of this book is to help you start prepping for a full month of living on only your prepping supplies. That may seem like a daunting goal at first, but once you achieve it you'll never want to go back to a state of being unprepared.

If you're like many people today, you carry hardly any cash, or at least not very much. Many people were advised years ago to not carry cash out of fear of being robbed. In fact, according to a survey of over 2,300 Americans by Vouchercloud, 57 percent said they "never" carry cash—compared to only 10 percent who say they "always" have cash on them.

Regardless of the reason, financial transactions are just one more way that we are addicted to the electrical grid.

Credit and debit cards are rapidly replacing cash transactions, and digital payments are just as quickly replacing credit and debit cards. From online banking and PayPal to digital currencies and smartphone transactions, cash is disappearing, and fast.

When I was a kid it was hard to make money. Now it's hard to even find it.

Electronic banking and payments may make modern transactions convenient, but in a disaster it can create—well—a disaster. During a widespread power outage or a bank holiday, how would you access your own money? If you can't access or use it, is it really even yours?

When electricity fails, credit card terminals, ATM machines and bank computers do not work. Therefore, from the corner grocer to hotels and gas stations, all transactions will be cash only. If you rely primarily on payment cards or online banking for purchasing, you'll be at great risk during any significant emergency.

Even if there is no widespread emergency, think back to the list of most likely disasters I included in chapter two. Remember #1 on the list? It was job loss—and the best way to be prepared for a job loss is to have at least a three-month supply of liquid assets on hand.

BEING FINANCIALLY PREPARED FOR A JOB LOSS

I often say that the best time to plant a tree was 40 years ago. Likewise, if you find yourself suddenly without a steady paycheck, you'll realize that the best time to have put aside a safety net was in prior years while you were earning money.

A job loss—even a sudden one—will sting a lot less if you have your finances in order when it happens. This can save you from having to go into debt or being forced to quickly take a non-desirable job, rather than taking the time to find the job that's right for you.

One great way to be prepared for not having a paycheck is to be in the habit of generating multiple streams of income.

For those of you who have even a small piece of land, I covered tons of ways you can do this in the book *How to Make Money Homesteading*, which also includes lots of ideas that don't require any land at all, such as writing, consulting, and affiliate marketing.

Beyond the ideas in that book there are numerous tasks you may be able to do to earn money locally, depending on your skills. Almost everyone can do something, such as walking dogs, mowing lawns, providing transportation, selling crafts, etc.

If you're working now, take the time to create a personal austerity budget—one that will allow you to determine the minimum you need to cover basic expenses, such as rent or mortgage payments, utilities, food, transportation and health insurance. If you're not saving enough and are concerned about your financial security, go ahead and implement your austerity budget now while you are working so that you create a safety net.

To show you what I mean, here's a sample austerity budget:

Item	**Normal Monthly $**	**Austerity Monthly $**
Rent	$900	$500
Health Care	$400	$400
Cell Phone	$200	$45

Item	Normal Monthly $	Austerity Monthly $
Electric	$150	$100
Groceries	$400	$300
Eating Out	$200	$50
Entertainment	$200	$50
Clothes/Other Goods	$300	$100
Internet	$150	$0
Total	**$2,900**	**$1,545**

This is a simplistic budget, of course, but illustrates what I mean by enacting an austerity budget. I've done much of this myself, actually. For example, I kept my iPhone but switched from AT&T to StraightTalk, lowering my phone bill from $200 per month to $45. Likewise, most of the other expenses, other than health care, can be reduced or eliminated by a motivated person. While you may not be able to reduce your rent expense, I reduced it above to show the effect of bringing in a roommate, should you have that option.

Keep in mind that if you find yourself without a paycheck you don't necessarily have to make as much money as you earned before. Rather, there are many things you can do now (while you have a paycheck) to reduce your needed expenses later. For instance,

- learn to grow much of your own food,
- get completely out of debt,
- sell unused items you have lying around and put the cash aside,
- learn to hunt and fish for free protein,
- implement your austerity budget to eliminate non-

essential expenses, such as cell phone plans, satellite television, dining out, and unnecessary travel, and

- start a self-employed business on the side before you need the income from it.

Few people would associate reducing expenses and implementing austerity measures as actually improving their quality of life. Rather, the terms more often bring to mind suffering and having to go without.

However, a growing number of people are using that precise formula to do something unheard of in decades past—to retire at a very early age, often in their 30s and 40s.

An excellent book to introduce you to this concept, if you're intrigued, is *Early Retirement Extreme: A Philosophical and Practical Guide to Financial Independence*. The book's author, Jacob Lund Fisker, also operates an excellent blog and forum where aspiring retirees discuss methods to reduce expenses and retire very early, at earlyretirementextreme.com.

CHECKPOINT—CASH RESERVE

- Make an austerity budget.
- Determine how much cash you want to have on hand.
- If you have trouble achieving your cash target, implement your austerity budget until you meet your goal.
- Write a summary of your current state, what you would like it to become and how you will get there.
- **Example**: I need to have $500 in cash in tens and twenties. I will implement my austerity budget until

I achieve that in cash. In addition, I'll continue the austerity budget until I have at least three months of savings stored in the bank.

STORING MONEY AND BARTER ITEMS

The most liquid of all assets in our society is, of course, cash, but don't overlook other items that you could quickly liquidate if you suffer a job loss and needed to raise cash. Those may include sellable items such as livestock, vehicles, tools, collectibles, and other items you could sell quickly if necessary.

If you're able to accumulate and store cash personally (not in a financial institution), don't just keep one hundred dollar bills. Instead, I recommend keeping twenty-dollar bills, as those will be easy to use for most transactions and don't draw the attention that one hundred dollar bills do.

How much cash you should keep will depend on several factors, such as how far you may have to travel in an evacuation, how well stocked you are in terms of supplies, and the number of people you're financially responsible for. Will you have to pay cash for fuel and a hotel when you evacuate, or do you already live at your rural retreat?

If you can establish an emergency cash fund all at once then that's one less thing to worry about. If you can't, just build it up over time. Start with $50 or $100 and add a set amount each month. Soon you'll have a supply of cash to get you through a crisis. Ideally your emergency stash will grow to several hundred dollars in small bills, at a minimum.

If you're not comfortable storing cash—get comfortable.

There are plenty of places to store money, and you'll need money to survive.

So where do you keep the cash? Here are some considerations:

- First, understand that whatever you do with your own money, you do at your own risk. There is no FDIC coverage for money stolen from your mattress.
- Many preppers advocate hiding money within the walls of a house. Personally, I've never liked this idea because if you experience a house fire you'll suffer two disasters—the loss of the house and the loss of your cash. But to each his own.
- In my view a better choice is a very well made, fire-resistant safe bolted to a concrete floor.
- Another good choice is to hide money in a fireproof container within a fireproof building, such as a metal storage building not adjacent to your house.
- Store some cash (not all of it) under a spare tire in your car's trunk.
- Always keep an adequate amount of cash on you or in your car emergency kit should you become stranded. Don't simply rely on credit cards or ATMs working, and it will do you no good if all your cash is at home.
- If you're concerned about theft, create a small "throwaway" safe you keep on your nightstand. Since 95 percent of robberies are smash-and-grab, thieves will quickly grab the safe thinking they have your valuables. Instead, hide your cash in another location.
- If you have a basement or crawl space, you could hide

cash in a bag within a fireproof container and store it there. Use desiccants as needed to keep the money from absorbing moisture.

- There are many other places you can hide money, such as burying it in the garden, in potted plants on the patio, geocaching, etc., but you run the risk of forgetting where you put it. Keep it safe, keep it close, and tell only those you trust with your life where it is—in case something happens to you.

I suppose it is notable that in a book about prepping I've yet to recommend storing precious metals—namely gold and silver. The reason is because the point of this book is to *start prepping now* for short-term disasters, which I've defined as a month or less.

While I definitely believe that "hard" assets are a critical element of any prepared person's holdings, I don't believe it is critical for you to own precious metals simply to survive for a few weeks. Having the ability to satisfy your needs for food, water, shelter, sanitation, and defense is much more important, and more easily attainable.

Clearly, keeping a reserve of cash for emergencies is something I strongly recommend, but your cash is *only* going to help if the stores are open and stocked, which they may not be in the event of many disasters. This is why you not only need cash stored, you also need food, water, and other items covered in this book.

In a widespread and long-lasting emergency, stores will likely have empty shelves, meaning your neighbors will be just as much in need as you are. They may not be willing to part with any of their life-sustaining supplies for your money, but may trade for something they can use in the moment. That's another reason why it's a good idea to store more than what you actually need for yourself.

Then again, in a widespread and long-lasting emergency, stores will likely have empty shelves, meaning your neighbors will be just as hungry as you are. In that event, don't expect anyone to be willing to part with their supplies for your money. After all, they have to feed their kids, too. Their goods are essential for survival, so unless you have something they need (medical supplies?) they likely will not be willing to trade.

Certain items in particular tend to be highly sought after during emergencies and may be particularly useful for barter. These include

- batteries,
- matches,
- candles,
- fuel,
- toilet paper,
- ammunition,
- alcohol/cigarettes, and
- hand tools, etc.

Basically anything of value can be used for barter. It all depends on your neighbors' wants and needs. You can also barter your services. Perhaps you're good at performing first aid or at butchering an animal, for example. These are skills you can practice now.

Craigslist has a category on barter items. Why not take a look at it and practice bartering for items that will help increase your preparedness?

CHECKPOINT—BARTER ITEMS

- Determine what items of value you have available for barter.
- List other items you could stock up on that would be of value in a barter situation.
- Write a summary of your current state, what you would like it to become and how you will get there.
- **Example**: I've identified several potential barter items, including a case of wine, two cases of toilet paper, and several unopened packs of batteries. I'm going to stock up on extra #10 cans of rice and beans for barter as well, as well as extra packs of calcium hypochlorite for water purification. Each of those items are inexpensive now, but should have high barter value in a disaster.

PAYING BILLS

As we've covered, many emergency scenarios may make it difficult for you to travel and access cash, so you'll want to have enough cash at your location to last you the month, or however long you wish to prepare for.

However, you also need some ability to pay certain expenses *without making physical contact.*

For example, consider a widespread pandemic where you don't want to go to the bank, grocery store, or utility office. In that example, the grid (and Internet) will likely be up so you'll have the ability to pay bills online. This means you must have money in the bank and an Internet connection. Or you can rely on the old-fashioned approach of paying bills from your mailbox without venturing into public. Just use your checkbook and keep a book of Forever Stamps.

PHYSICAL STORAGE OF DOCUMENTS AND DATA

Food and water are among the first categories new preppers think about, of course, but many new to prepping overlook the importance of having secure storage of critical documents and data.

Whether it's a death in the family or a widespread disaster, every family needs a plan to survive, cope, and recover from the hardship. By creating an emergency binder, your family will be better prepared to endure unexpected adversity with minimal disruption.

It is important to keep your important **personal documents and data** in a fireproof safe. Here is a list of documents you'll want to ensure are readily available in your emergency binder:

- Birth Certificates for every family member (preferably certified copies).
- Marriage and divorce records.
- Adoption records.
- Vehicle titles.
- Vehicle registration documents for each vehicle.
- Photocopies of deeds (preferably certified copies).
- Social Security cards for every family member.
- Stock certificates, savings bonds, and brokerage ac-

count statements.

- Proof of loans made and debts owed.
- Wills and Power of Attorney.
- Pension documents.
- Immunization records.
- Previous year's income tax return.
- Photocopies of household bills (utilities, phone, cable/satellite, Internet, etc).
- Partnership and corporate operating agreements.
- Photocopies of your credit cards.
- Video/photos of your home possessions.
- Backup of important photos/videos/digital files.
- Photocopies of all insurance policies (home, auto, health, life).
- Photocopies of photo ID cards (driver license, concealed carry permits, student ID, etc.) for each family member.
- Lists of all prescription medications you and other family members take regularly.
- Names, addresses, and phone numbers for your primary physician, dentist, attorney, and insurance agent(s).
- Information relating to bank accounts (name of bank, account numbers).

- Photocopies of all credit cards (front and back).
- Photocopies of property ownership records (deed, land contract, mortgage paperwork, etc.).
- Other (personal and unique documents, such as DD Form 214, Diplomas, etc).

Also, be sure to keep current photos and descriptions of your pets to help others identify them in case you and your pets become separated, and to prove they are yours.

ELECTRONIC STORAGE OF DATA

If you haven't already done this, begin by purchasing a thumb or flash drive. These small USB devices can store an enormous amount of information. Be sure to password protect the contents of the flash drive given the sensitive nature of the contents.

Once the USB drive is ready, either download or use a scanner to make electronic copies of all of the documents previously listed, as well as any other data you feel may be necessary to have. Be sure the file types being saved are common ones that can probably be opened on any computer, such as DOC, PDF, or JPG files.

You should also consider copying important family photos and videos to an external drive. These drives have become remarkably affordable in recent years.

Given how large photo and video files can be, I use the Western Data Passport Ultra Portable external drive, which stores one terabyte of data for less than $60. We routinely sync our new home movies and pictures to the device, along with pertinent personal data, and store it in a fireproof safe.

As you download your documents and files to the drives, print hard copies if you don't have them on paper already. By having two sets of the information, you'll be in a better position to ensure at least one set will be available to you.

Okay, so now you have two complete sets of all this critically important data. Where do you store it?

My recommendation is to make an arrangement with a trusted family member or very close friend. Ask them to keep one copy of your files in a fireproof safe and you can do the same for them. That way if your fireproof safe is destroyed (in a flood, for instance), you'll have a backup at their house.

Another secure way to store the documents is to rent a safe deposit box.

For smaller and more critical documents, such as identification and medical records, consider carrying the password-protected USB stick in your everyday carry kit, if you have one—more on this later.

HOME INVENTORY

You've probably heard the recommendation before to inventory your household contents—but have you done it? It's a simple thing to do with modern technology.

Just take an hour or less to narrate a video of each room's contents and place the file in a safe place. Storage on a cloud server is a good example, but also having a backup on a thumb drive makes sense.

Having secure copies of this information outside the home but readily accessible can be crucial in expediting insurance claims and cutting through the inevitable red tape that comes along with disaster recovery.

CHECKPOINT—DATA & DOCUMENTS

- Copy and assemble your physical documents.
- Decide where you would store your documents and cash.
- Write a summary of your current state, what you would like it to become and how you will get there.
- **Example**: I've photocopied my important documents and given a copy to my sister to store in her gun safe. I have a digital copy of the documents and a video of the house contents on a thumb drive in my survival bag.

FINAL THOUGHTS ON MONEY & DOCUMENTS

Some of the suggestions in this chapter are among the easiest in this book to implement. Thumb drives are inexpensive and pretty much everyone has access to a printer/copier and a video camera. Therefore, it's easy to copy and scan personal documents, video house contents, and store everything securely on a USB drive or in a safe.

Yet, few people actually complete these tasks.

If you find yourself financially challenged to address some of the other areas in this book, at least take care of the ones you can accomplish.

I realize that it may take you a while to save an emergency fund and to get comfortable storing cash. But many people fail to do anything when they can't do one thing. Don't let the perfect be the enemy of the good. Do what you can, now, to be better prepared.

Don't forget your pets

When you're copying your documents, remember to have current photos and descriptions of your pets in case you need to help others identify them if you and your pets become separated. Photos and descriptions will also serve to prove the animals are yours.

Finally, don't forget to have information on their feeding schedules, medical conditions, behavior problems, and the name and telephone number of your veterinarian in case you need to board your pets or place them in someone else's care.

It takes more than crossing your fingers to be prepared. Stock up, make a kit, form a plan and share it with your family.

Tim Young

CHAPTER TEN

PREPAREDNESS STEP 8: TRAVEL SECURITY

It may surprise you that I've devoted a step to travel security on the 10-Step Path to Preparedness, but this is a testament to the fact that we spend so much time away from our homes.

The fact that the phrase "bug-out bag" is used so frequently underscores the importance of travel security, for none of us know for certain *when* a disaster will occur or *where* we will be when it does. After all, what if the disaster strikes while you're at work or on vacation?

Snowed Out Atlanta

Consider what happened to Atlanta area motorists on January 28, 2014. Even though every local news station told residents to expect up to two inches of snow, few took the warning seriously.

By nightfall just over two inches of snow had fallen—an amount that seems comical at first. However, the surreal landscape on roadways more resembled a nightmarish scene from *The Walking Dead* than any comedy.

"HELP NEEDED!!! INFANT IN CAR!"

> "My friend has a 3-month old baby and has been trying to get home. 9-1-1 is slammed! She has no more water to mix bottles and barely any

> gas. If anyone can help with a 4X4 please respond!"

This was a frantic message posted to *SnowedOutAtlanta*, a Facebook page hastily established to aid stranded motorists. There were many other messages, just as urgent:

> "I'm eight months pregnant and have my 3-year-old with me," Katie Horne wrote. "We've been in the car for over 12 hours. Is anyone near on the road that might happen to have any food or some water?"

By mid-afternoon most people in Atlanta panicked, raced out of work and clogged the roads. Rebekah Cole was among them, leaving her Atlanta-area office mid-afternoon, only to find herself halfway home TEN hours later at 1:00 a.m., stuck on the interstate. Snarled in an eerie gridlock of unmoving cars, many of them abandoned on the highway, she prepared to spend the night in her car as temperatures plunged into the teens.

> "If I get gasoline, I'll turn the heater on and keep the windows cracked a little bit," Cole told herself. "I'll be all right."

In the distance she could see a gas station, but it was swamped with hundreds of cars ahead of her trying desperately to inch forward for the last of the gas.

Then the fuel light in Rebekah's car went on.

From minor inconveniences to real disasters, problems can happen at anytime. Being prepared means being ready *without notice*, for any unexpected event that jeopardizes your safety.

EVERYDAY CARRY ITEMS

In a moment I'll discuss survival bags, but first I'd like to cover items I carry on me at all times. This will help you understand why I make the choices I do for my survival bags and automotive kits.

As you expand your own survival mindset, you'll want to go beyond a survival bag, which could easily be forgotten or stolen. In that event, each of us has certain tools that we never want to be without.

Those items are referred to as "everyday carry items."

Those new to prepping may consider everyday carry items to be bulky and uncomfortable, but think about what the "socially acceptable" everyday carry items are now. Many women carry purses filled with all kinds of things they may or may not need. What if these items were replaced with some prepping items? Businesspeople often have briefcases; children take backpacks to school and many city dwellers do the same when traveling by subway.

Finding room may be as simple as changing the type of bag you carry and taking advantage of small, compact and multi-purpose prepping items available. I'll show you some that I use, but there are many more options that you can search for.

Now—here is my list of "everyday carry" items I have on me at all times when I'm out:

- Wallet with cash and identification (driver's license and concealed carry permit).
- Leatherman Surge multi-tool.
- Firearm & holster.
- Collapsible self-defense baton (when I can't carry a firearm).

- Pepper spray.
- Reading glasses.
- QuickClot bandage (since I carry a firearm).
- Bic lighter, even though I don't smoke.
- Magnesium fire starter.
- Compass.
- Small mirror.
- Survival whistle.
- Tactical ballpoint pen.
- Folding magnifying glass.
- Password-protected USB drive with important data.
- Lib balm.
- Pocket first aid kit in Altoids can (potable aqua tabs, aspirin, band aids, sanitizing wipes).
- Duct tape (wrapped around a plastic card).
- Fishing kit wrapped in paracord.
- Analog wristwatch.
- LED flashlight.
- Smartphone (with camera, GPS, emergency and first aid apps, compass, etc.).
- Miniature locking pliers.
- Stanley pocket screwdriver.

This seems like a lot to carry, especially without a purse, but I rely on a couple of clothing options that make it a snap.

One option I use is 5.11 brand TacLite Pro pants, shorts and tactical shirts. They all feature lots of spacious pockets and are extremely well made.

Another option I depend on, particularly in warmer months, is to wear a lightweight photographer's vest. On Amazon the one I use is called a Plainclothes Concealed Carry Vest. It has lots of pockets and in the summer I wear it over a T-Shirt, which makes it easy to conceal a firearm if I'm carrying outside the waistband.

By carrying a few important survival items such as these I'm always at a basic level of preparedness, and it means that I have more room in my survival bag for other items.

CHECKPOINT—EVERYDAY CARRY

- Consider what items you may need to carry at all times. In other words, if you were trapped in a high-rise elevator, what items might you wish to have on you? Ask yourself more "what ifs" like that.
- Think about how you could carry those items.
- Make a list and organize a plan.

TRAVEL SAFETY

My daughter loves to visit grandma and grandpa. Getting there requires a several-hour drive through very long stretches of economically depressed and sparsely populated countryside. The kind of stretches where there is no phone signal, no street lights, and no gas stations or stores for dozens of miles.

It's a very bad area to break down anytime, but especially at night with a wife and young daughter. That's precisely why I take travel security and preparedness so seriously.

Think about the last time you planned a car trip. Did you study a map to see if your route took you through any isolated areas where you might be unable to reach help? Or did you essentially hop in the car and go, blindly assuming that the car would get to its destination without incident? That gas stations or roadside assistance would simply be available if needed? That the GPS would work flawlessly?

Most people do make those assumptions, I'm afraid. They know they have a spare tire in the vehicle, but they've never changed a tire before on the roadside. If the vehicle overheats or they run out of fuel, they assume they can call a friend, family, or roadside assistance.

But what if they can't—what if *you* can't?

Perhaps your car breaks down in the evening, but you have a phone signal and are able to make a call. Roadside assistance can be there within two hours, they tell you, but what if a couple of rowdy guys pull over and offer to help? What if they insist? Can you trust them? Did you bring any personal protection with you?

What if it isn't nighttime, but rather a blistering summer day when you break down on an isolated stretch, with no shade around and two children in tow? Do you have water? Do you have shelter? Do you have a first aid kit?

Most people don't consider these possibilities, so if you too are like this, you're normal.

Stop being normal. Start being prepared.

To ensure our safety, we keep "get-home bags" in our vehicles at all times. You can also think of them as "survival bags."

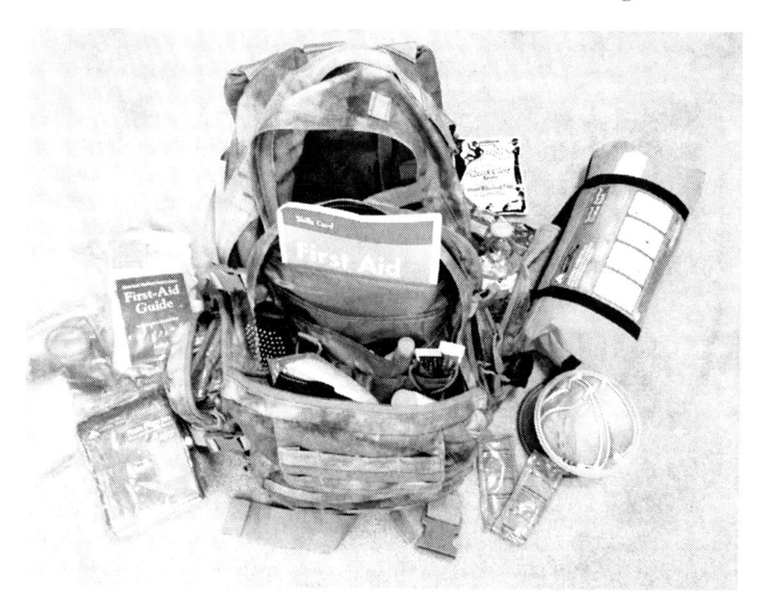

One of the author's survival bags

The idea behind carrying survival bags is that we may be an hour or more away by car from our preps or a safe destination when the grid goes down, cars stop, or we simply have a breakdown in an isolated area.

I know, I know: your normalcy bias tells you this is not likely, but as you can probably tell by now, I think it's prudent to be prepared. So don't judge me; just accept that I take the responsibility of being prepared for my family seriously, and I want to encourage you to do the same.

By the way, a survival bag like this would have been very handy in the Atlanta snowstorm and during the stranding of the Nevada and Oregon families mentioned in the introduction, so it *does* happen!

When a widespread disaster happens, what will separate our vehicle from every vehicle on the road? The answer is my everyday carry items and our survival bag, fully stocked to sustain us for days.

What should go into a survival bag? Let's take a look.

SURVIVAL BAGS

A survival bag, sometimes called a bug-out bag, is a backpack loaded with emergency survival supplies. It is particularly useful for those who live in densely populated areas who feel they may need to "bug out" if something bad happens, such as in the event of an evacuation.

I prefer the term survival bag to bug-out bag since the goal may not be to "bug out," but rather to get home or survive a roadside emergency.

Many merchants sell pre-made survival bags, but I suggest you make your own. Some purchased kits are poorly assembled and none are customized for your specific needs. Also, when you carefully consider what you need and pack the bag yourself, you'll know exactly where everything is stored and why you included it.

Before I share recommendations on building your survival bag, keep in mind that the purpose of this book is to help you start prepping and get you prepared for a disaster that may impact you for up to one month. For that scenario the most likely reason you might have to bug out would be evacuation of some sort, such as getting out of a major city, evading a wildfire, flood, or hurricane.

Survival bags are generally designed to sustain a person for the 72 hours or less that it takes them to find more permanent refuge, such as with friends, family, or a shelter.

If I were guiding you on how to prepare for a longer-term potential disaster, such as an EMP or grid-down scenario, your survival bag would be assembled to support life for many months of wilderness living. Therefore, it would emphasize fishing supplies, snares, foraging, gardening, butchering, natural shelters, primitive cooking, and more.

If you're interested in "upping" your preparedness plan on your own, customize the basic survival bag I describe below to your own desires.

Remember, we're talking about YOUR safety and survival, so don't let anyone tell you what should be in *your* survival bag. You know your environment, your needs, and your family's capabilities better than anyone. You decide, but here are some ideas to get you thinking:

Eight Guidelines For All Survival Bags

1: Be discreet. Sure, there are lots of cool-looking military backpacks that are quite functional, but they *scream* preparedness and may make you a target if others are in desperate need of supplies. Go for something tattered and mundane that you can find at a yard sale, or consider covering the bag.

2: Go lightweight. Big, heavy bags are—big and heavy. You may have to hike 20 to 30 miles in a day, so opt for lightweight bags and contents. Our choice for a bag was a Condor 3-Day Assault

Pack that I bought on Amazon, but ultimately the bag should be packed to match your physical capabilities. If you must include more than you can comfortably carry, consider getting a foldable golf-bag pull cart or a jogging baby stroller and strap your contents to it. Both are easy to pull and designed to go over uneven terrain.

3: Compartmentalize. Keep common items in pockets together, such as fire starting supplies together, water purification together, first aid together, tools together, etc.

4: Think multi-use. The classic example is to opt for a multi-tool instead that can replace several items, such as a blade knife, scissors, screwdrivers, can openers, etc.

5: Be waterproof. A simple way to keep your bag and contents protected from the elements is to use a contractor trash bag over it, which also serves as an inconspicuous way to conceal it. Also, be sure to store matches, socks, and other items within your backpack in quart and gallon Ziploc bags if you want them to remain dry. It's a good idea to toss oxygen absorbers into each bag if you live in a wet or humid area. If you have the ability, a better idea may be to vacuum seal these items, which will not only keep them dry but compress them as well.

6: Collect containers. Tic-tac containers, Altoids cans and so on make excellent storage compartments for many items that will go in your bag, and can even be lined to serve as "EMP proof" devices for USB drives, for example. Collect them now.

7: Stay current. Each season it's necessary to do a quick inventory check of your bag and adjust the contents as needed. This means changing to appropriate clothing for the season, checking expiration dates, etc.

8: Train. Once your bag is packed, practice taking it on an extended hike. That way you'll know what it feels like to transport the weight and see how far you can go, so you can adjust as necessary. Also, don't just train in good weather. Take a hike in rain and/or snow if possible to see what you may need to be ready for. It's a good idea to do this each season when you update your bag.

14 Categories To Include in Survival Bags

1. Water. Water is priority number one. Given how heavy it is, carry only one day's supply of actual water per person. Beyond that, include LifeStraws and the ability to disinfect water.

2. Food. Because the bag is only designed to support you for two to three days, you shouldn't overemphasize food. Include long-lasting comfort food such as a small jar of peanut butter and some crackers, protein bars, jerky, etc. Include plastic utensils for the peanut butter and remember to replace the crackers as they hit their expiration dates. Also, toss in a book or descriptive pictures you've copied of local edible plants and what they look like in each season.

3. Shelter. Tents are heavy. Instead, rely on lightweight tarps and space blankets. For a short-term crisis you'll likely be able to find existing shelter structures, so the main goal is to keep warm and dry. The exception to this is if there are two or more adults traveling together. In that case it may be convenient for one to carry a tent.

4. Clothing. The contents here depend on where you are (warm/cool climate) and what time of year it is (summer/winter). You'll need to adjust your bag throughout the year to ensure you have adequate clothing to keep you warm or cool, dry and protected from the elements. Don't forget to pack sunscreen, hats, gloves, hand warmers, etc.

5. Personal protection. Many prepared citizens carry a firearm at all times, and they are indispensable tools if you're comfortable with them. Otherwise, you can rely on pepper spray, batons, knives, and the like. Remember—if you're in a situation where you need a survival bag, others will need one too. Only they didn't plan ahead like you did so they'll be eyeing yours. You must protect it.

6. Tools. Carry a multi-tool such as the Leatherman Surge that I carry on my belt at all times, meaning I don't need one for my survival bag. These tools include screwdrivers, multiple knives, scissors, wire cutters, files, rulers and more. Also, don't forget to throw in paracord, which can come in handy for rigging shelters and numerous other needs, as well as zip ties and duct tape. Finally, include a lightweight and bright flashlight. At the time of this writing you can purchase a Refun 2000 Lumen Handheld Flashlight on Amazon for less than $15. It requires three AAA batteries, which I recommend you store in their original package separate from the flashlight. When you need to bug out, simply insert the fresh batteries.

7. Fire. Include a disposable lighter, magnesium sparking tool and a small container of 100 percent cotton balls that you smeared with petroleum jelly. To start a fire, remove a pinch of the soaked cotton ball, fluff it out and ignite with a lighter or magnesium tool.

8. First aid. Following the recommendations from chapter seven, assemble an effective medical kit for your bag based on your family's needs. Don't forget to include your first aid manual.

9. Sanitation. A small pack of baby wipes can help keep you clean and can be used for field showers as well as for emergency toilet paper.

10. Navigation & communication. Carry a physical map of your area that has important destinations already marked, such as campgrounds, friends, family, etc. If you're planning for potentially longer-term scenarios, map out water sources and potential fishing and hunting grounds. Also, it's a good idea to pack a hand-crank weather radio and be sure to include your cell phone with a solar charger.

11. Money. Include cash and coins in your bag, enough to last you 72 hours. If you're worried about theft, keep bills in the shoes on your feet.

12. Personal documents. When you assemble the emergency document binder covered in the previous chapter, make digital copies of all your documents, store on a password-protected thumb drive and include in your survival bag. That way you'll have access to important documents if you lose your physical copies in a natural disaster.

13. Personal needs. Pack whatever you need for your family's personal needs, such as inhalers, diabetic medicine, feminine napkins and so on. Feminine napkins are useful not only for their intended purpose, but they can also help stop bleeding from puncture wounds.

14. Pet needs. If you have one or more pets, you'll likely want them to bug out with you. They can eat jerky if you pack enough, but you'll need leashes and anything else they require, as well as ensuring they have access to water.

Many people store their survival bags inside their home near the door, so they can grab it at a moment's notice. That makes sense in many situations, as you'll see later in this chapter when

I cover evacuations, and disaster forces you to vacate your home.

Of course, the opposite could happen just as easily—you're away from home and want to get back to your preps.

This issue of getting home to where my preps are is the topic that I fret over the most, and I encourage you to think long and hard about it as well. After all, if you do follow the advice in this book and achieve short-term preparedness at home, what good is it if your family is in the car, on vacation eight hours away? What will you then do?

You can't possibly take all your preps with you, so it's a dilemma. That's why we have a fully stocked survival bag in each vehicle, *at all times*. If you don't want to keep it in the car, create a place for it beside your door. Then, get in the habit of grabbing it every time you leave and returning it to the same spot when you come back home.

In addition, we take advantage of the extra space a vehicle provides by stocking other items that could, as hard as it is to imagine, be the difference between life and death.

CAR EMERGENCY KITS

I'd like to encourage you to ALWAYS keep your fuel tank at least half full! This is critical in case evacuation is needed, but also in case you need to use your car for emergency heating, cooling or to power an inverter as described earlier.

Also, pre-set one of the buttons on your car radio to the emergency weather station, and tune in when needed.

Of course, I have my everyday carry items on me, so those are items I don't necessarily have in a survival bag—though I do carry back-ups for some items. And, by always wearing good walking shoes on our feet (rather than heels, crocs, or flip-flops) we save the space of carrying an extra pair in the car.

Now, in addition to the contents of our survival bags, here are items we also have in the car at all times. I encourage you to do the same and to not assume that your vehicle will always work smoothly or that help is just minutes away.

- Heavy-duty (4 gauge) jumper cables.
- Tire pressure gauge.
- Automotive tool kit.
- Flares.
- Fix-a-Flat (Pressurized tire inflator and sealer).
- Emergency triangle.
- Gasoline siphon.
- Windshield deicer and scraper, in season.
- Towing strap.
- Fire extinguisher (A-B-C type).
- Gerber folding spade with serrated blade.
- Rags.
- Spare tire, tire tool, and jack (and knowledge of how to use them!).
- Car safety hammer, window breaker and seatbelt cutter clipped to the driver's seatbelt.
- A junior dome tent (in case we're stuck somewhere

rural at night).

- Smart phones with chargers.
- A separate backpack with a sleeping bag, space blanket, duct tape, change of clothes, long underwear and heavy socks (in season), work gloves, poncho, jacket and/or sweater, bandana, gloves and hat.
- Medical supplies including a first aid kit, quick clot, Israeli trauma bandages and N95 masks (in the event of pandemic), capsules, tweezers, lip balm & sunscreen, soap, baby wipes, towel, insect repellent, scissors, prescription drugs, toothbrush, and personal items..
- A portable Kelly Kettle stove.
- Shelf-stable food including tubes of peanut butter and DATREX 3,600 calorie emergency bars (not the tastiest thing but it will sustain you, and when you need it, it's there). We also include eating utensils, a pocket knife, knife sharpener, cooking kit w/cup, can opener and moist towelettes.
- Bottled water for 24 hours, a LifeStraw and water purification tablets.
- A stroller for the young'un at all times (or to carry our bags).
- An umbrella (for rain and sun protection).
- As a concealed carry permit holder I always have a firearm, and am careful to consider neighboring state reciprocity laws before crossing state boundaries. We also carry pepper spray and collapsible batons.

I'll admit that the above list looks like a lot to carry and you may wonder if you'll have any room left for the kids. As I mentioned, you'll have your own list, but the goal is *for you to have what you need to reach safety without a vehicle.* Most of these supplies are important, just as the spare tire is. You can probably find room. Just use all those empty spaces, such as where the spare tire is stored, rear cup holders, door storage compartments, under seats, seatback pockets, glove compartments and consoles, above visors, and so on.

I strongly suggest you create your own survival bag and car emergency kit for your family. Otherwise, you run the risk of being in a remote location protected only by the *hope* that your vehicle will continue to run normally.

CHECKPOINT—SURVIVAL BAGS & CAR EMERGENCY KITS

- Do you have a stocked survival or bug-out bag?
- Imagine that you become stranded at dusk on a rural roadside. What items will you want to have in your car for comfort, protection, and survival?
- Determine if you have a stocked first aid kit in your car and know how to use it.
- List the car emergency items you have in your vehicle. Do you have jumper cables, tools, de-icers, flares, flashlights, etc?
- Write a summary of your current state, what you would like it to become and how you will get there.
- **Example**: I have a stocked survival bag, but I need to check the contents more frequently so I can change items with the seasons. I don't have a set of jumper

cables, so I will purchase a heavy duty set for my car immediately. It scares me to imagine being stranded and having to confront someone, but I'm not about to carry a gun. I'll be sure to have pepper spray and a stun gun in the car at all times. I've never carried food or water in the car, but I'll get a small cooler in the back to hold some bottled water, jerky, peanut butter, and crackers.

A WORD ON SENSELESS TRAGEDIES

It's sad to read stories of tragedies that could have been prevented with basic preparedness supplies. A recent news report painfully illustrated this point once again. Here was the headline:

> **Grandfather and his dog found dead after getting locked inside his Corvette by a dead battery in 90F heat in crowded parking lot**
>
> A 72-year-old Texas grandfather became locked inside his 2007 Chevy Corvette after the car's battery cable became loose. Police officials said that they found footprints all over the inside of vehicle, showing the man had desperately tried to escape. The loose battery cable meant that even if he had honked the horn in the parking lot, no one would have heard him. The elderly man was also unable to call for help because he had left his phone in the restaurant.

The problem is that many of us dismiss these stories as something that couldn't happen to us. Yet they do happen. There are several lessons we can take from this tragedy so that we can each become better prepared to live and prevent needless suffering.

- Having a car safety hammer inside the vehicle provides the ability to break car windows. Virtually everyone overlooks this device. Get one that has a seatbelt cutter as well—and make sure every occupant knows where it is kept.
- We would all be well served to carefully study our vehicle manuals. It's a credit to technology and engineering that we are free to take operation of our vehicles so casually. Put gas in, occasionally change the oil, and the car just keeps going. Few of us actually read the manual and know how to change fluids, change tires, or fully understand the vehicle's features. Corvettes like the one in this story have an emergency handle on the floorboard to force open the doors if electronics fail, but many drivers are unaware of the feature. Before your next trip take out the manual and read it in detail. It could save your life.
- While even the best prepper may accidentally leave her mobile phone behind, her everyday carry items would likely include a firearm or multi-tool, either of which could be used to escape a car.

The news is replete with these tragic stories. The best we can do is to learn from them and take simple, affordable measures so that we can all be prepared to live and not be victims.

EVACUATION

You've probably never had to evacuate. If you have, then you've likely already started prepping pretty seriously, as you know first-hand the hardship that evacuations can bring.

There are a number of reasons that you may have to evacuate, and many come with virtually no warning. Those include raging wildfires, earthquakes, nuclear accidents, and chemical leaks. Even those that offer more notice, such as hurricanes, generally affect a larger area, so evacuation can be very slow, if not impossible.

The best hope is to be prepared to leave quickly should the need arise.

When a 2003 firestorm ravaged San Diego County, terrified residents were shaken out of their beds by police loudspeakers:

"If you can hear my voice, you are in immediate danger and must evacuate now."

Four years later in 2007, a series of wildfires again torched southern California, forcing *one million* people to evacuate. The fires destroyed 1,500 homes. That's a lot of people and a lot of homes, so it can happen to any of us.

There are many reasons why a last-second evacuation may be ordered, from hurricanes and floods to wild fires, chemical spills, terrorist attacks, viral outbreaks and nuclear accidents. Now is the time for you to know evacuation routes for your area. This is also an excellent opportunity to teach your children about both maps and the importance of planning.

What would you take if you had to evacuate? Many people don't think clearly during times of stress and panic, so a well thought out Bug-Out List that you create before an emergency will make things much easier.

Your Bug-Out List should have items *listed by priority*, because you don't know how much time you'll have to evacuate. After all, not all evacuations can be planned in advance the way most hurricane evacuations can. You may get a knock on the door in the middle of the night and be told you have a few minutes to escape a wildfire or chemical leak.

Even in the event of hurricanes, people are often not prepared to evacuate until it's too late, as gridlocked motorists discovered when they tried to evacuate Houston during Hurricane Rita. If an evacuation order is given, take it seriously and leave immediately.

Hopefully you'll never experience an evacuation, but take the time NOW to learn from others' experiences. Most will tell you that the better prepared you are, the fewer regrets you'll have.

Consider these things when planning your evacuation bug-out list:

- Who is important to you?
- What is important to you?
- What would you need if you were gone for a few days to a week?
- If everything in your home were to disappear, what items would make starting over easier?

HOW TO ORGANIZE YOUR EVACUATION BUG-OUT LIST

- List items that are essential to you. It may be a list of random thoughts at first. That's fine, just keep refining and narrowing it down.
- Then, prioritize your items into two groups. Items at the top of your list are your absolute essential items to

grab if you had only five minutes to evacuate. Items on the second list are those you would add if you had 30 minutes or more to prepare.

- Make three columns for your list. One column is for the item itself, one column is for the floor of the house the item is located on, and the final column is for the specific location on that floor. This will prevent you from running around frantically during a stressful situation only to miss necessary items in the same area you already visited.
- Ask your family members what things in their rooms are the most important and irreplaceable to them, and pay particular attention to what children say. You may not be able to take everything they ask for, but in a stressful situation, the stuffed Peppa Pig may be just the thing they need to stay calm.
- Make multiple copies of the list and post it around your house. Make sure it is visible to all family members.
- Conduct a practice evacuation with your family and use your list. You should assign different family members to gather different items.

The first time you do this activity, you'll likely realize that you're not as prepared as you'd like to be. Just keep practicing and, hopefully, this approach will make a very traumatic situation just a little less stressful for you and your family.

The following pages illustrate example evacuation bug-out lists to get you started.

EXAMPLE EMERGENCY EVACUATION BUG-OUT LISTS

5 Minutes Warning

Evacuation - 5 minutes notice	**Floor**	**Location**
72 Hour Survival Bag	**Main**	**Closet**
Wallet or purse	Main	Kitchen
Baby bag with extra food/formula	Main	Kitchen
Car/house keys	Main	Kitchen
Cell phone, charger & car charger	Main	Kitchen
Pets	Main	L.R.
Pet food, & leash (for walking dog or shelters)	Main	Kitchen
Prescriptions & eyeglasses	Main	B.R.
Cash/Purse/Wallet	Main	B.R.
Extra contacts or glasses	Main	B.R.
Medical devices (wheel chair, etc.)	Main	B.R.
Laptop	UP	Office
Irreplaceable photos	UP	Office
External hard drive	UP	Office
Child stroller	Main	Garage
Adequate shoes for weather	Main	Garage
Important documents (already in safe)	Main	Garage
Pet carrier	Main	Garage
Coat or jacket if necessary		

Family Reconnect Plan (next chapter)	
Plan A Meeting Place	Home
Plan B Meeting Place	School
Plan C Meeting Place	Police Dpt.
Plan C Meeting Place	Grandma's

EXAMPLE EMERGENCY EVACUATION BUG-OUT LISTS

30 Minutes Warning

(add these items to 5-minute list)	Floor	Location
Emergency supply bucket of food	Main	Garage
Sleeping bag	Main	Garage
Tent	Main	Garage
Lanterns	Main	Garage
Crank radio	Main	Garage
Camp stove	Main	Garage
Journals	UP	Office
Box of family pictures	UP	Office
Desktop computer	UP	Office

In addition to your evacuation bug-out list, you need an **Evacuation Action List**. These are the critical things you need to do, if possible, just before leaving.

For example,

- close and lock all windows,
- unplug appliances,
- turn off/set to minimum A/C and heat,
- turn off main water into home,
- turn off breakers for non-essential power, and
- turn off gas, if relevant.

Post an instruction list for how to do each of these actions in the event you're not home and the task falls to another family member, babysitter, etc.

If there's a chance the weather will worsen or flooding may occur but authorities haven't yet advised an immediate evacuation and you're *sure* you have time, take steps to protect your home and belongings.

- Ensure safety for your pets in accordance with your personal planning. This may mean taking them inside the home or preparing to transport them with you.
- Load your survival bag and emergency supplies into your vehicle, or put them by the door if you have to leave on foot. In some disaster situations, such as a tsunami, it could be better to leave by foot so you can quickly get to higher ground that roads may not reach.
- Bring items indoors, such as lawn furniture, trash cans, children's toys, garden equipment, hanging plants, and any other objects that may blow away or break windows.
- Look for potential hazards that could blow or break off and fly around in strong winds.
- Turn off all utilities coming into the house.
- If strong winds are expected, cover the outside of all the windows of your home. Use shutters that are rated to provide significant protection from windblown debris, or pre-fit plywood coverings over all windows.
- If flooding is expected, use sandbags to keep water away from your home. If you live in a flood area, be sure you have enough sand, burlap, or plastic bags, shovels on hand at all times.

If you're forced to evacuate, it is critical that your family has an agreed-upon communications plan and plan to reconnect. This need is addressed in the next chapter.

CHECKPOINT—EVACUATION

- List potential causes in your area that could cause a sudden evacuation. For example, nuclear, chemical, wildfire, hurricane, tsunami, etc.
- Complete both a 5-minute and 30-minute bug-out list, following the sample in this chapter.
- Communicate the plan with your family and rehearse it once a year.
- Write a summary of your current state, what you would like it to become and how you will get there.

FINAL THOUGHTS ON TRAVEL SECURITY

I talked a lot about getting home and bugging out in this chapter, but "bug out to where?" you ask.

Many folks aspire to have a bug-out cabin or retreat, but let's face it: that's expensive, isn't it? Not many people can afford that. I mean, how many people do you know who have a fully stocked bug-out retreat?

If you live in the city or suburbs but don't have a bug-out retreat, perhaps you know someone who does live in a secure rural setting accessible to you. If so, try to establish an arrangement so you can go there during a disaster.

If you do, just remember that city and rural folks tend to have different values and perspectives, and live at a very different pace, so if you go this route it is important that you forge a relationship with them based on your need for mutual survival and preparation. That means visiting with them from time to time to help them out, and perhaps inviting them to your area to show them the city while times are good, if they're interested.

Sometimes it is challenging for folks of different backgrounds to get along. Ironically, nowhere is this more true than within families, but there will be no room for that in a crisis. Build those relationships now as part of your preparedness planning.

Of course, my recommendation is to live at a rural retreat full-time if possible, which is what I do.

Don't forget your pets

Keep a sturdy leash, harness, and carrier to transport pets safely. A carrier should be large enough for the animal to stand comfortably, turn around, and lie down. Your pet may have to stay in the carrier for several hours.

Remember pet toys and the pet's bed, if you can easily take them, to minimize their stress.

The single biggest problem in communication is the illusion that it has taken place.

George Bernard Shaw

CHAPTER ELEVEN

PREPAREDNESS STEP 9: COMMUNICATION

The thought of facing a widespread emergency and not knowing where your family members are or if they are safe is chilling. Equally frightening for your relatives is their not knowing if *you're* safe when they learn of a disaster in your area. So, if disaster strikes and modern-day communications fail, what will you do?

If you think this is unlikely, think again.

During Hurricane Katrina, Cingular Wireless (now AT&T) lost more than 85 percent of its cell towers in the affected area.

Likewise, consider these sobering findings from a study published by Consumer Reports regarding Superstorm Sandy in 2012:

- 75 percent of people lost power for at least one day, with a median of *seven* days.
- 73 percent lost Internet and television service.

During a time when so many people were cold, hungry, injured and terrified, the desperate victims of Superstorm Sandy had no way to reach loved ones or be reached themselves.

Clearly each prepper needs to have a plan for communicating when disaster strikes, but it doesn't take a hurricane or blizzard to take out cell towers—nor does it take something as dramatic as an

EMP or terrorist attack. Heavy snowfall, torrential rains, or even software glitches can interrupt telephone service.

In fact, a mundane software bug *crashed* the AT&T long-distance network in 1990, resulting in 60 million call attempts unable to get through. Heck, even when towers are operational, they often can't handle the added calls on high traffic days such as Mother's Day.

These examples emphasize the critical need of having a foolproof family emergency communication plan that works even when conventional routes don't.

FAMILY EMERGENCY COMMUNICATION PLAN

The objective of a family emergency communication plan is to make sure your family members will know what to do and how to reconnect when the time comes. The objective is to clearly articulate a plan so that, in a sudden crisis, the family goes into a preprogrammed mode and executes the agreed-upon plan.

A disaster can happen at any time and without warning, such as when you're asleep or when you and your loved ones are separated by work and school. Since your family may not be together when a disaster strikes it is important to plan *and rehearse* in advance.

Your family emergency communication plan should describe *clearly* what you want to happen during an emergency:

- What is expected of each member during an emergency?
- How should each person attempt to communicate (call, text, radio, email, etc.) and in what order?
- When will the plan be activated?

- Will everyone head home if communications are down or should they meet at a rendezvous point?
- Will you go to a secure area or stay put, and how will you decide?
- How will everyone get to a safe place? For example, if communications go down, dad will pick up the kids from school while mom drives home from her job, etc.
- How you will get back together, and what you will do in different situations?
- If something happens and you can't find one another, what will you do?

As you create your emergency plan, have a family discussion to determine who will be the out-of-area contact person, and where you'll meet away from your home—both in the neighborhood and within your town. These will be your rendezvous points should you have to evacuate.

After that discussion, write down important information on a family communication plan card and be sure each family member keeps a *physical* copy.

Don't forget to inquire about emergency plans at places where your family spends time, such as work, daycare and school, churches, YMCA, sports events, etc. If no plans exist, consider *volunteering to help create one.* This is also a great way to find others who are interested in prepping and to gently encourage people to become more prepared.

Talk to community leaders, your colleagues, neighbors and members of faith or civic organizations about how you can work together in the event of an emergency. You'll be better prepared to safely reunite your family and loved ones during an emergency if you communicate with others in advance.

What follows are templates to get you started, but take the time to customize each for your needs.

EMERGENCY CONTACT INFORMATION

Out of Town Contact
Name:
Home #:
Cell #:
Address:
Email:
Facebook:
Twitter:

Meeting Places
Home Address:
Neighborhood Location:
Local Location:
Regional Location:
Distant Location:

Health Insurance Information
Medical Provider:
Phone #:
Policy #:

Medical Contacts
Family Doctor:
Phone #:
Pediatrician #:
Phone #:
Dentist:
Phone #:
Veterinarian:
Phone #:

Home Insurance Information (in case of house fire, flood, etc.)
Insurance Provider:
Phone #:
Policy #:

Family Member #1 (example, and create for each family member)
Name: Dad
Cell #: 212.555.1212
DOB: 12/13/1963
Medical Information: High blood pressure, **Allergic to penicillin**

Work Information
Workplace:
Phone #:
Address:
Facebook:
Twitter:
Evacuation Location:

School Information
School:
Phone #:
Address:
Facebook:
Twitter:
Evacuation Location:

FAMILY COMMUNICATIONS TREE

Before you contemplate the different tools and methods available for communication, I recommend you write out a family communications tree for your family.

What's a family communications tree? Here's a visual of what it looks like:

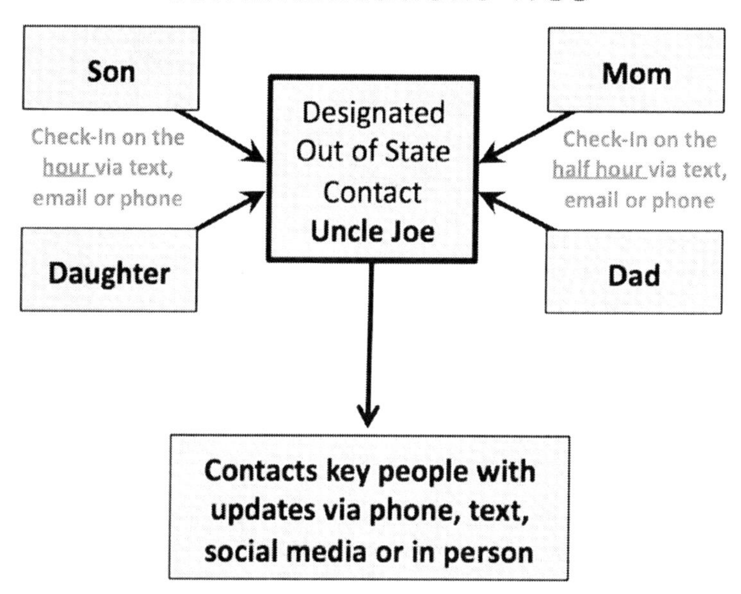

I call this a *communications* tree rather than a *phone* tree because there are many methods you may have available to communicate in an emergency. Most likely you won't be limited to a single method, and, unless you're facing a TEOTWAWKI scenario, I suspect there will be viable communication methods other than ham radio.

Here's how it works:

- Designate an out-of-state contact who is far away. In other words, if you live near the state border don't simply choose someone on the other side of the border and think you have a distant "out-of-state" contact. Rather, make sure they're far enough away to not likely be affected by the same disaster you're facing.

- Make a plan in advance so each family member knows to contact the designee. The designee is the central source of information for the family, so he or she must be available and reliable.
- Provide at least three methods of reaching the designee. For example, text message, telephone, email, Twitter, Facebook, radio, etc.
- Ideally, each family member will have a laminated emergency contact card that they carry in their wallet, just in case they are unable to access the contact information on their mobile phone.
- Be sure every member understands how to use each method and that the designee does as well.
- Have the parents contact the designee on the half hour and the children on the hour. For example, the children may call at 8:00 a.m., 9:00 a.m. and so on, while the parents check in at 9:30 a.m., 10:30 a.m. and so on. Of course you can customize this to your specific needs.
- During each check-in the designee can report to each family member that the others are okay. This will calm family members and reduce each member's need to contact others directly, thereby freeing up lines for emergency use.
- The designee is responsible for updating key people, who will likely include relatives but may include doctors, employers, and more. Inform the designee of

how often he or she should update others. Daily may be a good choice for many situations.

It's important that the designee be located a good distance away—at least a long-distance call. Local and long-distance calls work on different circuits. When local circuits are overloaded, you often will still be able to make long-distance calls.

CHECKPOINT—FAMILY COMMUNICATIONS

- Create your emergency communication plan by gathering emergency contact information for all family members, along with the physical places they are likely to spend time.
- Choose your meeting places.
- Create your family communications tree and gain cooperation from an our-of-state contact.
- Make sure everyone knows the family's emergency communications plan and laminate a physical copy for all family members to see.
- **Example**: I have a list of all emergency contact information and my grandfather is the designated out-of-state relative. We have planned that, in an emergency, mom will pick up the kids at school and they are to wait there (or at the school's evacuation location) for her. Dad will travel home from work. If we are not able to go to our home, we will go to our town's YMCA to meet. IF we cannot communicate with each other directly we are each supposed to check in with grandpa—kids on the hour and parents on the half hour.

Now that you have created your family communications tree, let's look at the communication options available to you, in order of likelihood.

SMARTPHONES

When preppers discuss communications, they most often sing the praises of being an amateur radio operator, or ham. That's a fine solution and I don't want to dismiss its value, but for many people, ham radio can be confusing and expensive, especially to new preppers.

The fact is that very few people are or aspire to become amateur radio operators. Therefore, as my goal is to inspire you to start prepping now and become prepared for up to 30 days, I'd like to prioritize more commonly used methods to help you achieve your communications objectives.

Many preppers are hesitant to rely on technological contraptions, particularly Smartphones. The argument is that if the grid goes down, the devices and networks, which rely on electricity, will be of no value. That may be true, but again let me refer you to the list of likely disasters in chapter two, most of which will result in some utilities still being available. Therefore, a Smartphone is a wise choice for many reasons.

- Texting—Fast and simple, texting is a great way to let others know you're okay and to make sure they are too. Text messages require far less bandwidth than phone calls. Since texts operate on a separate data network from cell phones, they work even as you hear the ominous "all circuits are busy" recording when attempting a call.

- Email—If you can access a WIFI, you'll likely be able to successfully send email. Email servers are distributed globally so it is unlikely they will all be down. Just be sure to have at least one cloud-based email account, such as Gmail, rather than relying on private accounts, such as an email server at your place of employment, which may be inaccessible in a local emergency. Free WIFI access is often found around restaurants, coffee shops, libraries and other public places.
- Powerful apps—There are a number of fantastic apps that can help you to both communicate with loved ones as well as provide critical knowledge to help you survive. I've provided a comprehensive list of Smartphone apps in Chapter Twelve.
- Telephone service—While cell towers do go down and circuits become jammed, you may get through, particularly during the middle of the night.
- Social media—Tweet your status or post an update on Facebook. YouTube videos require more bandwidth so I don't recommend that channel for emergency communication. Social media is similar to email in that it is hosted on a network of global servers, which offers excellent redundancy and fault tolerance. It's easy to post Facebook or Twitter messages to your family members as a backup to the backup in case of an emergency. Just make sure family members follow your social media updates.

Earlier I shared a story of the 2014 Atlanta snowstorm as a disaster that affected thousands of motorists. While those motorists could have been better prepared, particularly in terms of having survival bags, the fact is they were able to use Smartphones and social media sites to communicate. The same is often true with wildfires, tornadoes, riots and other emergency situations, proving the value that Smartphones have as a communication option.

TEXT, DON'T TALK!

- Unless you're in danger, send a text. Texts may have an easier time getting through than phone calls, and you don't want to tie up phone lines needed by emergency workers.

CELL PHONES AND LANDLINES

Cell phones are differentiated from Smartphones mainly because they lack the ability to use powerful applications that can be of valuable aid in an emergency. Like landlines, their chief virtue is in establishing vocal contact, although virtually all cell phones, unlike landline phones, can also be used for text messages as well.

Most people have abandoned corded landlines in favor of mobile devices. Yet having a landline could be a valuable insurance policy in an emergency, since traditional landlines don't require electricity and will work during a power outage.

Landlines have proven to be more reliable during natural disasters than mobile devices, for when a natural disaster or an emergency occurs, many people use their cellphones and bombard cell towers, reducing the chances of getting through. True, you may not be able to reach loved ones in an affected area unless they too have landlines, but you can at least report extended power outages and reach emergency personnel.

If you don't have a landline yourself, perhaps you can get to a library or public building and find one that you can use.

PAY PHONES

Yes, pay phone booths still exist, though it has likely been many years since you've used one, if ever. Still, if you can't complete a mobile phone connection and don't have a landline, a pay phone could help you complete your call. Most are based on landlines, which are inherently reliable.

Look for pay phones in common locations, such as at gas stations and convenience stores as well as at transportation hubs such as bus depots, train stations and airports.

TWO-WAY RADIO

A two-way radio is simply a radio that can both transmit and receive. Also referred to as "walkie-talkies," two-way radios are typically equipped with a "Push-To-Talk" PTT button to activate the transmitter.

These are a fine choice for very local communication, but it depends on where you are. If you are in a city jam-packed with skyscrapers, your reception will be a couple of blocks or less. The same is true, I'm afraid, if you're in a thick forest, since two-way radios work best with line of sight.

Line of sight doesn't mean the radios need to see each other, but rather means that the signals travel in straight lines and don't bounce off the atmosphere the way AM radio signals do, potentially traveling for hundreds of miles. Two-way radio signals can bounce off surfaces and penetrate objects, but they travel in straight lines. Due to the earth's curvature, these straight lines eventually go off into the atmosphere after a few miles.

To help you determine the maximum range you're likely to achieve without a fixed antenna, here's a quick formula for calculating a distance to the horizon:

- Measure the height from the ground that the antenna of your handheld two-way radio will be. For me, this is six feet.
- Determine the square root of that height. In my example, the square root of six feet is 2.45 feet.
- Multiple that result by 1.22 to determine the distance from the horizon. For me, the handheld radio will be 2.99 miles from the horizon (2.45 X 1.22), at which point the radio wave will continue straight into the atmosphere.

Given this example, I would only expect line of sight communication for approximately three miles, but this assumes that the receiving radio is lying flat on the ground. That's an unrealistic assumption. If the other person's radio were also held six feet off the ground, this would extend the communication to approximately six miles.

This suggests that the maximum range of these devices is four to six miles, and that is on flat ground with no obstructions.

If you're in the city, get to the top of a building or parking garage and you'll greatly increase the potential range. As for me in the country, the best I can do without an antenna is climb up into my deer stand.

You may see two-way radios advertised that can transmit 30 miles or more—but don't expect those results unless you plan on raising an antenna quite high.

Still, two-way radios have a couple of nice benefits:

- Instant communication - Two-way radios have a "Push-To-Talk" (PTT) mike as well as voice-operated (VOX) transmitters. Once the button is pressed, the user can instantly convey his/her messages rather than waiting for the device to ring and then for someone to answer.
- Hands-free—The VOX feature makes them ideal "hands free" systems for hunters, cyclists, hikers or motorcyclists. The audio can be heard directly through earphones, keeping the transmissions private from those around you.
- Group communication - Another great feature of two-way radio in emergencies is its ability to allow group communication. In this way, one user can talk to many at the same time, eliminating the need to repeat communication to each recipient.
- Reliable—Two-way radios are not dependent on operational telephone service or electricity. Just keep them fully charged and have an alternate method of charging should electricity be unavailable. Many two-way radios operate simply on AA batteries.
- Licensing—An FCC license is not required to transmit using a two-way radio as long as it is operated in the Family Radio Service (FRS) mode.
- Affordability—Most two-way radios are very inexpensive.

- Quasi-Private—Given that they use public airways, conversations on two-way radios cannot be considered completely private, but radio features such as unique privacy codes can filter out general transmission of conversations, making these more private than CB radios. Of course, the privacy of two-way transmissions will depend greatly on the population density of your surroundings. If you're in the city and lots of people have two-way radios, chances are you'll hear other conversations—and they will hear yours as well.

There are many two-way radio options to choose from, and it's good to have some even if it's only for your family outdoor adventures. We have a pair of the Motorola MH230R 23-mile range radios, just in case we can't use cell phones and need an alternative communication method. They work well for us, but we don't expect to be able to communicate more than a couple of miles with them.

CB RADIO

When we think of Citizen's Band (CB) radios, many of us recall movies or events from the 1970s when CB transmissions were all the rage. From "breaker breaker" to "10-4 good buddy" to "what's your 20?" much of the CB jargon has found a lasting place in our vocabulary.

However, there were no mobile phones in use during the 1970s when CBs became a great means of mobile communication. That's different today, of course, when many children seem to be born tethered to a Smartphone. Still, CB radios remain operational and are in use.

Like two-way radios, CB radio is a public, two-way personal radio service. They tend to be much more bulky to operate than two-way radios, and therefore are usually mounted in a vehicle or home. However, most CB radios connect to an external antenna, which generally provides a much greater range than that achieved by two-way radios.

During disasters, CB radios are considered by some to be superior to two-way radios because emergency responders monitor CB channel 9, designated by the FCC to be an emergency contact channel.

Like two-way radios, conversations using CB radios should not be assumed to be private, since the radio transmissions use public airwaves.

Unlike two-way radios, there will likely be many people monitoring one of the 40 channels within the 27 MHz band available to CB radios, including, as mentioned, channel 9. This makes it much more likely that you'll be able to reach others and learn important news and events.

CB Radio offers several benefits.

- Licensing—No license required to purchase or use.
- Monitoring—Easy to monitor what is happening on highways. For example, if you're stuck in a long line of bumper-to-bumper traffic on the highway and wondering if you should take an alternate route, a CB radio may be able to provide that answer more quickly (and more safely) than using a Smartphone. If you monitor the trucker channel (19), you'll hear frequent traffic updates.

- Dependability—CB radios can use a car's 12v electrical system.
- Affordability—Like two-way radios, CBs are relatively affordable, particularly when compared to amateur radio set-ups.

While there are some benefits to CB radio and it may be a good solution for some, I generally don't recommend them for someone interested in short-term preparedness. They have a place in long-term grid-down scenarios as they can be used as a communication platform for a local community, but since CBs have a slight learning curve anyway, I would advise a new prepper to ease into amateur radio instead.

AMATEUR RADIO (HAM)

Amateur radio, commonly known as ham radio, is a hobby enjoyed by several hundred thousand people in the U.S. and by millions of people worldwide. Amateur radio operators are often referred to simply as "hams."

Using ham radio, you can take mobile devices wherever you go and communicate from a mountain top, your home or while on a sunny beach, all without relying on the Internet or a cell phone network.

Amateur radio operation is a fun, lifelong hobby. When times are good and there are no emergencies, you can talk to other hams through one of several satellites in space, bounce signals off the moon and back to Earth or even talk to astronauts aboard the International Space Station (ISS)! Yes, you read that correctly. Some ISS crewmembers make random, unscheduled, amateur radio voice contacts with hams, and astronauts have contacted thousands of hams around the world. Hams can make direct contact with the ISS station when the crewmembers are working.

In times of trouble, amateur radio can also provide reliable communication during states of emergency since it works when other services fail.

For example, when Hurricane Andrew struck South Florida in 1992, it destroyed the utility grid, all cellular towers and antennas over hundreds of square miles. Only amateur radio, CB and pay phones with underground lines were effective at communicating with those outside the affected area.

And, during the 9/11 tragedy, New York City agencies used amateur radio to keep in touch with each other after their command center was destroyed.

Ham radio is the most regulated of the non-commercial communication services and can be the most expensive, but that's because it is the most versatile and powerful. However, before you can get on the air, you must be licensed and know the rules to operate legally.

There are three amateur license classes in the U.S. —Technician, General and Extra.

The easiest way into ham radio is via the Technician class license, which requires a written test based on a text available through many sources. A great and free resource to get you started with your training can be found online at www.hamradiolicenseexam.com

The Technician class license is the entry-level license. It requires passing an examination totaling 35 questions on radio theory, regulations and operating practices. The license grants access to all Amateur Radio frequencies above 30 megahertz, allowing these Technicians the ability to communicate locally and most often within North America.

The General class license grants some operating privileges on all Amateur Radio bands and all operating modes, and opens the door to worldwide communications. Earning the General class license requires passing a 35-question examination, and applicants must have previously passed the Technician examination.

If you have the energy, you can aspire to earn the Extra class license, which conveys all available U.S. Amateur Radio operating privileges on all bands and all modes. Earning it is more difficult and requires passing a thorough 50-question examination. Of course, Extra class licensees must also have passed all previous license class written examinations.

While amateur radio does require training and licensing, there is a great parallel benefit to it that you won't find with the previous communication methods I've described, which is the availability of support from local amateur radio clubs. You can find a list of clubs on the American Radio Relay League website at www.arrl.org/find-a-club. Who knows—you may even meet some folks there who share your values on preparedness!

The real value of amateur radio is tapping into the network of repeaters. By using a repeater, a ham radio can be used to reach halfway across the U.S.

As part of your family emergency communications plan, be sure that your family knows what frequencies you can be reached on, and what times you'll be listening. For example, every two hours at the top of the hour starting from 8:00 a.m.—6:00 p.m., etc. Also, be sure to carry a repeater guide for your area at all times rather than relying on the programming in the radio.

STAYING UP TO DATE

Beyond the need to communicate to others, it's critical that you have the ability to receive emergency information. Here are some items that can help you to stay informed:

- Police scanner—You can download a police scanner Smartphone app, but the safer choice is to monitor a physical police scanner. This is particularly important in areas of greater population density, where crime and civil unrest can escalate—and get out of control—quickly.
- Emergency alert system—Use your Smartphone or email to subscribe to emergency alert updates. As of this writing, I use the free Red Cross Emergency App on my Smartphone. It provides real-time updates on tornadoes, hurricanes, tsunamis, floods, storms and more, plus has a "Family Safe" function that allows me to instantly contact loved ones after alerts are issued in my area.
- Emergency hand crank weather radios—A one-way radio to receive ambient weather and alert systems is a useful communication tool to own. Preppers often prefer to own a hand crank emergency radio such as the Ambient Weather WR-099 solar hand-crank radio. This radio works via sunlight or hand cranking, and offers the confidence of knowing if severe weather is expected.

FINAL THOUGHTS ON COMMUNICATIONS

Communicating with your loved ones during an emergency doesn't need to be difficult or scary. With a little pre-planning, you can feel safe and secure even when others are panicking.

Here's a summary of my specific recommendations for new preppers:

- Go over different emergency scenarios with family members and write out a family emergency communication plan.
- Rehearse the plan by actually going out and simulating the scenarios that your family identifies in the plan. Then, go back and modify your plan based on each member's abilities, tendencies and on other factors you observe.
- Write down an emergency communication plan and a family communications tree, and give it to each concerned family member.
- Have a Smartphone (ideally) or a simple mobile phone with you at all times.
- If you don't already have one, consider installing a landline with a very basic phone plan, just for emergencies.
- Learn the location of pay phones in your area.
- Purchase inexpensive two-way radios for local, family use.
- Get to know your neighbors and their communication capabilities.

- If ham radio is new to you, identify and visit a local amateur radio club to introduce yourself to ham radio.
- Visit the www.hamradiolicenseexam.com website to learn what it takes to pass the amateur radio Technician license. The website is free to use.
- Keep mobile device batteries fully charged at all times.
- Consider purchasing an alternative energy charger for your mobile phones.
- Subscribe to text alerts for weather and disasters that may affect your area. If you have a Smartphone, consider using the Red Cross Emergency app.
- Ensure each family member has a printed and laminated emergency contact sheet, complete with emails, phone numbers, addresses and social media accounts.
- If you do have a landline and must evacuate, forward calls from your home phone to your mobile device so that loved ones can reach you.
- Program "In-Case-of-Emergency" contacts into your phones so that emergency personnel know whom to reach.

A human being should be able to change a diaper, plan an invasion, butcher a hog, design a building, write a sonnet, build a wall, set a bone, comfort the dying, take orders, give orders, cooperate, act alone, solve equations, analyze a problem, pitch manure, program a computer, cook a tasty meal, fight efficiently, die gallantly. Specialization is for insects.

Robert Heinlein

CHAPTER TWELVE

PREPAREDNESS STEP 10: SKILLS

Thus far we've covered many items that will help you become prepared for most emergencies. From flashlights to medical kits and, yes, even firearms, it's great to have preparedness "stuff."

But stuff will only get you so far.

There's a debate in the prepper community centered on whether it's better to have supplies or to have skills. I think few people would argue against the need to have skills, and I'll devote most of this chapter to how to acquire them. But the truth is, unless you are the most die-hard, superhuman survivalist, you need both—stuff and skills—to survive.

And let's be honest—if you were that superhuman survivalist, you wouldn't be reading this book. So *you* need both.

The prior pages have touched on the kind of "stuff" needed as well as where to acquire it. But what about skills? Without skills you're simply an armchair prepper.

If you haven't had many opportunities to butcher animals, preserve foods or build wilderness shelters, fear not. The knowledge is out there in many forms just waiting for you to discover it.

This chapter includes a number of resources to help you acquire the skills and knowledge that will enable you to thrive during a crisis. I've included publications, online resources, Smartphone applications and training courses for your consideration. I'll also discuss the pros and cons of prepper groups since teaming with others is one way to expand your skillset. However, I'd like to begin with a more old-fashioned way to acquire skills: books.

A library of practical books is one of the most important preps you can have. The reason is simple—it is far more important to have skills than stuff, and absent hands on experience, books provide a deep well of expertise from which you can draw wisdom as needed.

Just as all preppers aspire to have a deep larder, you too should aspire to have a deep library of *printed* books. Kindle and eBooks are fine, but may not be there for you in certain scenarios. It would be awful if your deep library of electronic books were inaccessible, as it could be during some electromagnetic disasters.

When you purchase print books on Amazon you often find the publisher granted the right for you to have a free or low-cost Kindle copy. This allows you to have both digital and physical copies available, giving you both convenience and critical knowledge at your fingertips.

THE DEEP LIBRARY

I'm including a very comprehensive list of book recommendations for you to consider and have divided this list into two sections.

The first is what I consider to be the Top 10 Primary Preparedness Books. While you may need to wait to procure many of the Secondary Preparedness Books, I encourage you to get and read the top 10 as soon as you can.

Of course, this list assumes that you have purchased and will follow most of the preparedness recommendations in *this* book, so no book of mine is in the top 10.

Top 10 Primary Preparedness Books

1. *The Encyclopedia of Country Living*, by Carla Emery
2. *SAS Survival Guide 2E: For any climate, for any situation*, by John "Lofty" Wiseman
3. *LDS Preparedness Manual*, by Todd Assay
4. *Where There Is No Doctor: A Village Health Care Handbook*
5. *Food Storage for Self-Sufficiency & Survival*, by Angela Paskett
6. *All New Square Foot Gardening*, by Mel Bartholomew
7. *Basic Butchering of Livestock & Game*, by John J. Mettler
8. *The Forager's Harvest: A Guide to Identifying, Harvesting, and Preparing Edible Wild Plants*, by Samuel Thayer
9. *The Herbal Medicine-Maker's Handbook: A Home Manual*, by James Green
10. *Prepper's Home Defense: Security Strategies to Protect Your Family* by Any Means Necessary, by Jim Cobb

A detailed list of secondary preparedness books is included at the end of this book in Appendix III.

HELPFUL WEBSITES/PUBLICATIONS

- National Center for Disaster Preparedness—Located at ncdp.columbia.edu, the National Center for Disaster Preparedness at the Earth Institute works to understand and improve the nation's capacity to prepare for, respond to, and recover from disasters.
- Self-Sufficient Man—selfsufficientman.com is the author's website. It provides useful information and products related to living a prepared, self-sufficient life. Also, all suggested products mentioned in this book along with updated recommendations are listed at selfsufficientman.com/recommendations.
- Prepping/Preparedness Blogs – Due to the increasing interest in self-sufficiency and preparedness, there are a large number of blogs to choose from. Some are quite general in nature while others are focused on specific aspects of preparedness, such as food storage. There are far too many blogs for me to list here, so my recommendation is for you to do a web search for blogs in the area you are interested. If you need help getting started, SurvivalPulse.com created a list of the top 50 prepping and surviving blogs based on Alexa and Google page rank, and other criteria. You can find the list at survivalpulse.com/top-50-survival-blogs/.
- *Backwoods Home Magazine.* An excellent resource for independent, self-reliant living.
- PrepperShowUSA.com—PrepperShowsUSA is the #1 resource for survivalist shows, self-reliance shows,

and prepper shows across the United States. Find complete and accurate information about upcoming prepper show dates, schedules, show locations, venues, ticket pricing, promoter and exhibitor information and much more.

- Armstrong Economics—If you want to understand what's happening in global economics, Martin Armstrong's blog is required daily reading.
- *Pick Your Own*—All about home canning, freezing and making jams and jellies.

ONLINE DISASTER RESOURCES

There are many more preparedness resources than I list on this page, but these are ones I recommend you visit.

- FEMA—The Federal Emergency Management Association provides extensive information about preparing for and recovering from numerous types of emergencies and disasters at fema.gov.
- American Red Cross—The redcross.org site has useful information on preparing for emergencies and more.
- Ready.gov—Ready.gov is designed to educate and empower Americans to prepare for and respond to emergencies including natural and man-made disasters. Provides information to help people prepare for and respond to emergencies. Offers advice on assembling a simple emergency supply kit and creating a family emergency plan. Furnishes links to state and

local emergency management services.

- Centers for Disease Control—The emergency.cdc.gov website provides real-time information about public health issues, emergency preparedness, travel safety and many other important items for a prepared family. CDC also provides weekly updates of influenza outbreaks and publications that discuss infectious diseases.
- Community Emergency Response Teams—Located on the FEMA website, CERT educates people about disaster preparedness for hazards that may impact their area and trains them in basic disaster response skills, such as fire safety, light search and rescue, team organization, and disaster medical operations.
- Department of Homeland Security—The dhs.gov website provides a broad range of information relating to homeland security, including the national threat level, travel alerts and procedures, emergency preparation guidelines, immigration policies and border initiatives, school safety, and other security-related topics.
- Disability Preparedness—Red Cross has a useful document called Preparing for Disaster for People with Disabilities and other Special Needs. The PDF document can be found at tinyurl.com/9ycsspz
- Google Flu Trends—Track influenza outbreaks around much of the world. Compare current year's flu activity to previous years' at www.google.org/flutrends/

COURSES/TRAINING TO IMPROVE SURVIVAL SKILLS

Having food in your pantry and water on your shelves will buy you some time, but nothing will increase your self-confidence and odds of survival better than having real survival skills. There are several types of preparedness skills I recommend you pursue, and here are some places you can find them.

Homesteading Skills

I've had the opportunity to teach many classes on homesteading skills. An ever-growing number of folks are eager to learn the lost art of self-sufficiency, from butchering pigs, rabbits and chickens to making charcuterie, soap, and cheese to milking cows and handling livestock. Perhaps you're one of these folks.

If so, seek out a class and learn to butcher chickens, start a garden, make cheese, soap, or home medicines.

You can start by checking out localharvest.org for a list of farms in your area. See which ones allow visits and have volunteer days where you can help plant, butcher, and harvest. If you need to take a class in a certain locale, just do a web search for the class in your area—such as a chicken butchering class in Denver, for example.

A growing number of urban homesteading associations teach homesteading skills to city folks. Just do a web search in the city you're interested in to find them, but for starters, here are few:

Atlanta— thehomesteadatl.com

Denver— denverurbanhomesteading.com

Oakland— iuhoakland.com and farmcurious.com

Asheville— wildabundance.net

There are more homesteading groups and classes popping up all the time—just search for what you're interested in doing near your location.

If you have more time you could become an apprentice or intern. Check out wwoof.net for a list of worldwide volunteer opportunities on organic farms that range from a few days to a year or more.

Or, if you'd rather just take a working farm vacation, check out farmstayus.com for a list of participating farms.

To delve deeper into farming and homesteading attend some of the many conferences that are held each year. Examples include the Mother Earth News conferences, the Southern Sustainable Agriculture Working Group (SSAWG) annual conference, MidAmerica Homesteading Conference and numerous conferences sponsored by state organic farming groups.

You can dive into soap with the soap queen herself at brambleberry.com where you'll find lots of supplies and tutorials. From there you can seek out a local class if you'd like hands-on experience.

Likewise, learn to make cheese with the cheese queen, Ricki Carroll, at cheesemaking.com. Ricki's site has all the supplies and tutorials you need to make anything from simple cheese in your kitchen to award-winning cheese.

Also, don't overlook the importance of knowing how to sew. If you don't have a friend or family member who can teach you, there are numerous classes online that you can learn from.

If you're interested in curing meats and making charcuterie, start with Michael Ruhlman's excellent book, *Charcuterie: The Craft of Salting, Smoking and Curing.*

Finally, if you enjoy learning from videos, check out youtube.com/wranglerstar for lots of excellent prepping and homesteading tutorials.

Wilderness Skills

When my wife and I took our first vacation away from the farm several years ago we figured we'd go somewhere exotic. Instead we took a hide tanning class at a wilderness skills school where we learned that every animal has just enough brains to tan their own hide—well—except for buffaloes. And certain relatives of mine. The brain contains lecithin, an oil that serves as a natural tanning agent to lubricate the skin. Anyway, the class was a blast and ignited our fire to learn as much about survival skills as we could.

Perhaps you'd like to do the same.

Whether you want to learn to build shelters, make tools, or forage for your dinner, check out courses at MountainShepherd.com, CaliforniaSurvivalTraining.com, Earth School at lovetheearth.com or True North Wilderness Survival School at exploretruenorth.com. And there are many more schools and classes, so just do a web search for a class near you.

Firearms

The NRA offers great courses, but the best I've taken is a week-long course at Gunsite. There are also defensive firearm classes at frontsight.com and icetraining.us.

Many of these classes aren't just about how to shoot a firearm. They are about learning self-defense and tactical skills. This includes training for the survival mindset as I described in the Cooper Color Code System, but also tactical drills such as night shooting with flashlights, clearing houses, dealing with hostage situations and even real return-fire simulations, in more advanced classes.

Even if you're experienced with firearms, be sure to visit a local gun range at least once a quarter—shooting and gun handling is a perishable skill. Or, as Jeff Cooper put it so clearly:

> "Never assume that simply having a gun makes you a marksman. You are no more armed because you are wearing a pistol than you are a musician because you own a guitar."

Survival Medicine

You can actually learn a lot about survival medicine by watching videos. My favorite YouTube channels for survival medicine are youtube.com/ThePatriotNurse and youtube.com/drbonespodcast.

Videos and books are excellent study tools, but nothing beats hands-on experience.

For that, the Red Cross offers many basic first aid and CPR classes.

If you want to attempt surgery on that pig's foot I mentioned earlier, check out doomandbloom.net for a list of classes offered by Nurse Amy and Dr. Bones, respected authors of *The Survival Medicine Handbook*.

Wilderness Emergency Medical Services Institute offers wilderness emergency medical and first responder courses. You may have to travel to Ireland or Scotland to attend the course, but if you long to go, this could be just the excuse you're looking for.

If you have a few weeks and want thorough training, Soloschools.com may be just what you're looking for. Solo offers both home study and courses around the U.S. in wilderness first aid, first responder and EMT.

Don't overlook training in your own neighborhood. In many locations, EMTs offer training classes. If you don't see one listed, contact your local emergency medical technicians or the fire department to see if they can recommend someone. Also, you may be able to join your local volunteer fire department and earn training both in medical and fire suppression.

Finally, if you want to get serious about it, there are some excellent home-study herbal medicine classes, such as Sage Mountain's Science and Art of Herbalism.

Self-Defense

The goal of having exceptional situational awareness is not to engage in conflict, but rather avoid and escape it. Unfortunately, escape and avoidance may not always be possible. For those times, you'll want to be prepared to defend yourself.

Check out attackproof.com for a list of courses around the U.S. that will get you ready for any mugging or street fight. Classes are also held in the United Kingdom and Brazil.

Of course, there's always Krav Maga and mixed martial arts, which many preppers enjoy. Look for classes in your area—and get the whole family involved.

If you prefer to practice in the privacy of your own home, targetfocustraining.com offers a book and video course called *How to Survive the Most Critical 5 Seconds of Your Life*. The course has been featured on many major networks, including ABC News, CBS, Forbes and the BBC.

Communications

Visit the National Association for Amateur Radio website at arrl.org for a listing of amateur radio clubs in your area. Stop by for a meeting and stick your toe into the world of ham radio.

SMARTPHONE APPS FOR PREPPERS

We use technology today to find restaurants, check prices, and even to see if it's raining without looking outside. But many preppers have a love/hate relationship with digital devices.

The argument is that in the event of an EMP or similar event that takes down the power grid, these devices will be of no value. Therefore, diehard preppers emphasize the value of skills and the written word over any form of binary communication.

It's hard to argue with their logic—but who are we kidding? The genie is out of the bottle and, until such time when a solar flare fries the genie we're all going to use these devices. So why not embrace them for their convenience today while we make preparations to live without them if that time comes? That's why I recommended storing printed copies of books earlier in this chapter.

Here are a few apps that you may want to add to your digital preps. Most are available on multiple platforms, and many are free.

- Pocket First Aid & CPR: Developed by the American Heart Association, this app includes hundreds of pages of instructions on what to do in emergency situations, including dealing with choking, burns, CPR, seizures, diabetic emergencies, and more.
- SurvialGuide: This app is completely based on the U.S. Military Survival Manual FM 21-76. Survival skills are techniques meant to provide the basic necessities for human life: water, food, shelter, habitat, and the need to think straight, to signal for help, to navigate safely, to avoid unpleasant interactions with animals and plants, and for first aid.
- WISER: First responders and others exposed to emergency situations often have to deal with hazardous material. WISER helps those individuals by providing important information about hazardous substances and guidelines on what to do if you come into contact with such material.
- Wild Edibles Forage: Explore free edible plants in your backyard. Identify, cultivate, and prepare over 250 plants.
- The Bear Essentials: This app provides a complete blend of survival hands-on advice, interactive instruction manual, wilderness quiz, mini-games and adventure photo gallery, all rolled into one.
- SAS Survival Guide: Based on the bestselling book, this app provides you with a full guide to wilderness survival. Jam-packed with all the survival tools, you'll be equipped for any expedition to the outdoors with

this full-featured guide in your arsenal.

- ICE Standard: This emergency card app lists an individual's complete medical history, emergency contacts, insurance details, blood type, and more.
- Emergency Radio: More than just a police scanner, Emergency Radio has live police, fire, EMS, railroad, air traffic, NOAA weather, coast guard, and other emergency frequencies.
- ReUnite: This app enables a user to report and search for missing or found people reported to the People Locator (PL) Web site after a disaster.
- Proclivus. Updated every 15 minutes, Proclivius is a real-time news aggregator designed just for preppers. Thousands of articles & videos for DIY plans, checklists, medical issues, supplies, homesteading, off-grid living, survival, and just about anything else prepping related.
- Tiny Flashlight + LED: This app enables you to use a device's camera LED/flash/screen as a torch.
- Trimble Outdoors Navigator: View more than 68,000 topo maps in US and Canada, as well as aerial, terrain, and street maps.
- Knots Guide: Need to tie a knot? Knot Guide will teach you the ropes in 17 categories.
- Pet First Aid—A first aid manual for pets, from American Red Cross. It includes directions, procedures, text and short videos to show you how to take the necessary action step by step and in the right order.

- RepeaterBook: RepeaterBook enables every Ham to easily find repeaters across the World.
- Cargo Decoder: Cargo Decoder is your guide to what is in the truck or tanker next to you on the highway. Enter the four-digit number from the DOT placard to learn about a material.
- TruckerPath: Quickly find Truck Stops, Weigh Stations, Parking, Rest Areas, Truck Washes, Walmarts and Truck Services.
- Kindle eBook Reader: Carry a library in your pocket anywhere you go. Get access to more than 3,000,000 books including over 850,000 Kindle exclusive titles.

PREPPER GROUPS

Another decision preppers face is whether to go at it alone as a lone wolf or to partner with others and form a wolf pack. It's a tricky decision that appears to be easy to make, but generally isn't.

The old adage is that there is strength in numbers and, on the surface, there are many benefits of being part of a survival network.

While you surely have limited skills and supplies on your own, others could fill in the missing parts of your preparedness puzzle just as you perfectly fit a piece of theirs. It sounds too good to be true—and it often is.

Groups can either be intimate teams, comprising only very close friends and family members, or regional groups, comprising people you likely don't know but who have expressed an interest in preparedness.

You can find thousands of people across the U.S. who have already formed active regional groups on Meetup.com. Just go to tinyurl.com/prepgroups.

This may be a great way to meet people who may have skills you desire, such as those relating to food preservation, firearms and creating survival bags. By all means participate and learn from them what you can, and share what you know, if you'd like.

When you begin prepping you'll have a rush of excitement as a result of the skills and knowledge you're acquiring. Beware, because that "high" can wear off, causing you to lose momentum. Whether it's part of your intimate team or a regional group, one benefit of joining a group is that it can help you and others to defeat prepper apathy.

Be cautious, though, when contemplating inviting others to join your *intimate* prepper group. Sure, they can shoot and pressure-can food, but what do you know about them? Do you share the same life values? Can you trust them under stress?

Broadening your skillset is important, but *trust is everything* in a prepper group.

Therefore, learn from a broad group but only partner with those who, A) you can absolutely trust and B) who contribute a skill or supplies you need, and for whom you have a reciprocal value.

My ideal intimate team would be family members and close friends including doctors or medical professionals, mechanics, and those with tactical military experience. But they'd have to be people I've known for a while or am sure I can trust. To them, my wife and I can contribute much in the way of farming and hunting, food preservation and a detailed working knowledge of homesteading and natural remedies.

It's unfortunate and ironic that it's often difficult to form an intimate prepper team with the neighbors and family members closest to us. The reason is because, like many Americans, they simply don't share our preparedness mindset.

Rather than trying to convince them, I'd probably take an alternate approach and simply ask them to help me. If they haven't expressed an interest in learning about EMPs, financial collapse or otherwise, I wouldn't bring it up. Rather, I'd simply tell them I'm trying to take some precautions as a parent and could really use their help. Or, I may schedule a time to can food at an LDS Home Center and invite them to go along with me.

Over time, I hope they'll "see the light," but even if they don't, they'll understand everything I've been doing when a disaster hits, and can quickly become engaged in the team effort.

It's worth the effort, because the group can save money and increase its skillset much faster as a group than a lone wolf can alone.

A peculiarity I noticed when I moved to the countryside was that every landowner had his own hay equipment. It seemed wasteful to me that they each purchased and maintained a mower, fluffer, rake and baler, only to use the equipment twice a year, when his immediate neighbor had the same equipment. Granted, the conditions are "right" to cut and bale the hay at the same time for all of them, but there's enough leeway in the calendar to cooperate. They don't, however, because they're not "on the same team."

If you can create an intimate team with those you trust, you can avoid this. *They* can provide valuable medical supplies. *You* can provide pressure canners and Mason jars. Everyone wins.

CHECKPOINT—SKILLS

- Make a list of the skills you want to acquire, in priority order.
- Research ways you can acquire those skills.
- Make a concrete plan to allocate the time and money to acquire the skills.
- Write a summary of your current state, what you would like it to become and how you will get there.
- **Example**: I really want to learn how to defend my home and loved ones during a major disaster. I need hands-on training from a defense and firearms expert. I have researched classes and think ____ is the best one for me. The next class is four months away and I will begin saving now.

FINAL THOUGHTS ON SKILLS

Before I moved to the countryside and embraced a self-sufficient lifestyle, my wife and I had few of the skills we now value so much. Today, gardening, hunting, foraging, butchering, food preservation, shooting, herbal remedies, soap-, alcohol-, and cheese-making, beekeeping, carpentry and repairs are just a way of life.

These skills and a whole lot more are necessary to live a self-sufficient lifestyle, but if your goal is to simply prepare for a short-term crisis, I don't want to mislead you—most of them aren't necessary. After all, if your goal is to just have a two to four week supply of preps, you can't even get a garden started that quickly.

Still, there are some skills that I do recommend you acquire, even for short-term disasters. Make it a point to practice these 10 skills:

1. Know how to treat water.
2. Know basic first aid.
3. Practice staying warm without electricity.
4. Practice staying cool without electricity.
5. Practice cooking without electricity.
6. Develop your situational awareness and survival mind-set.
7. Know how to find or make shelter.
8. Practice multiple ways to start and control fire, even when wet.
9. Learn how to navigate without GPS.
10. Learn about and practice intermittent fasting at least once per month.

When you master these skills along with the skills and preps covered in the previous steps, and you can pat yourself on the back for completing the 10-Step Path to Preparedness!

Tragedies in the wake of disasters are too often the result of missed opportunities to prevent their occurrence.

Tim Young

CHAPTER THIRTEEN

ARE YOU READY?

It's easy to work yourself into a frenzy when it comes to prepping.

You can become consumed when a "reliable source" says that an asteroid will impact Earth in three months—though every reputable scientist denies it. Or perhaps a blogger you follow alarms you with a prediction of how little old North Korea will launch a nuclear EMP attack that will send America back to the Ice Age.

Your eyes grow wide when you hear a group fervently discuss a pending financial collapse, and suggest that you had best get what little money you have out of the banks—and put it into gold or silver.

Then you read of a recently discovered ancient prophecy that reveals the world will end very soon, so you wonder if it's worth preparing for anything.

For some people, hearing enough of these predictions means fear sets in. They often respond by buying all sorts of preps. Most people, though, simply respond by sticking their heads in the sand.

Unless you have achieved an advanced level of preparedness, there is no reason to worry about *what* event may impact your way of life. It simply doesn't matter.

Once you have achieved that level and have several months of preparations in all categories, then you can consider preparing in depth for a pandemic, multi-year grid failure or other possibility.

But you're not at that level.

You're likely new to prepping or looking to expand your preparations. You're looking to get started and that's why you chose this book.

At this stage you need to prepare for more "mundane" disasters such as personal tragedies, economic hardship and natural disasters. And I'm sorry to say, they're just as deadly and they happen very frequently.

Let me share one last survival story with you.

In January 2009, a severe ice storm crippled Oklahoma, Missouri, Arkansas, Illinois, Indiana, Ohio, West Virginia, and Kentucky. The storm killed 65 people and left two million people without power.

It was the worst natural disaster in Kentucky's history; 35 people died.

On a survival forum website, forum member KyFarmer shared a first-hand account of what his family experienced at his rural Kentucky homestead. As you read this, keep in mind that he was *already* a prepper. Even so, the storm made a lasting impression on him, and on me.

You can read the story at tinyurl.com/kentuckystorm. Let's see what we can learn from the story he tells in his own words:

> I've been through tornadoes, hurricanes, and blizzards—I've even survived an earthquake. But

I've NEVER seen anything like this. This was worse than most people had predicted or prepared for.

The storm came on Monday night with freezing rain. Freezing rain continued into Wednesday morning. Then the snow began. By the time all was said and done, I had about two and a half inches of ice and four inches of snow.

When I got home from work on Monday I cleaned the flue on the wood stove, moved seasoned firewood from the barn to the house, moved the generator to the house, took the Suburban to town to gas up, and stocked up on perishable provisions such as milk, lunch meat and so on.

Our power went down Tuesday morning about 9:00 am. No flicker, no drama, just dropped off.

That's never a good sign—but we fired up the generator and away we went. I have a 5,500-watt unit hooked up on a whole house relay switch. It'll run about 10-12 hours on four gallons. It runs everything in my house except the electric furnace (which I don't need because of the wood stove), and the cook top. If I shut everything off, I can run the water heater. Then I turn the water heater off and turn everything else back on so that my water pump works.

Temperatures never rose above 25°F until Saturday when it hit 40°F. The temperatures were a curse because they caused all the fields and side roads to turn to soup, making it hell for the power guys to get in and work.

There was a five county area around where I live with 100 percent power outage. The ice took down a trunk line and 10 substations.

My grandfather lives about 15 miles away, and it was Thursday before I could get to him. Keep in mind I've got a 3/4-ton suburban that's outfitted as a hunting buggy. Lifted, re-geared—the whole deal and it still took me that long.

The damage from falling trees was enormous. The "major" roads were difficult due to trees and lines down, but the secondary roads were impassable. I drove through fields at some points to get to my grandfather's house.

The biggest problem in my area proved to be fuel. There was ZERO gas, diesel and kerosene in my county and two neighboring counties for most of the week. If you got lucky and the station had enough power to run the pumps, there was a huge line of people.

Several counties lost water. When the substations went down the meager fuel supplies couldn't run the purification and pumping equipment long enough.

There was an entire county that the state essentially asked people to leave. It'll be at least a month before they get power.

At one point I had 14 people in my house, not counting the people that wandered through to take showers. I had a cousin in Missouri who bought eight generators at Lowes and drove them all night to get to us on Saturday. What our family didn't buy we took to town and sold for cost.

There were a lot of lessons learned.

Lesson 1—My wife and I learned that we're not nuts for being preppers. We're more convinced than ever that 12 months of supplies is a necessity.

Lesson 2—In light of the extended family, we have to reevaluate how much preps we need to accomplish our goals. We have to make some hard decisions on who and how much we are willing to help if it gets really bad. It will be very easy to be taken advantage of—we've got to put some serious thought into that. Problem is that the neighbors and friends know what we have stashed. If you have supplies and a plan, and your friends and family don't—it's best to consider that BEFORE there is a crisis.

Lesson 3—Pack Mentality is dangerous. A person is smart, but people are scared, panicky and not real bright. The fuel issues in town really drove that home. We had fights and lines and a shooting over fuel. In a crisis—if you're going into a situation where crowds are likely, ALWAYS *have an exit plan and a weapon.* Give serious consideration to whether the trip is a need or a want.

Lesson 4—Protect your stuff. People were stealing generators and siphoning gas out of neighbor's cars. It was nuts. I chained my generator to the support beams on my deck and joined the chains with a bolt rather than a lock so it couldn't be cut easily. You have to look objectively at what you have and ask, "Am I a target?" Is the placement of this item, the use of this item, or something I'm doing putting me at undue risk of attention?

Lesson 5—Decide NOW how far you're willing to go to defend your stuff and your home. After my issues with fuel, my wife and I had a talk about "what if." Don't get into a situation where you have to decide that under stress.

Lesson 6—Location REALLY matters. My house is in a really bad location and sits right on a paved road. If you have light and smoke when no one else does you're gonna have trouble. I had some trouble with people trying to steal fuel.

Lesson 7—DIVERSIFY! If you have substantial stores, spread them out in multiple locations. We're going to spread stuff out among several locations on the farm until we can get an isolated storage location built.

Lesson 8—STORE FUEL. If you don't have the capacity of a large tank like I do, have at least seven days of fuel in cans for your generator. Get a siphon hose and learn how to use it. Keep the tanks on your cars full and buy locking gas caps for them.

Lesson 9—Cell phones are not your friends. You don't realize how much you depend on them until you don't have them. We lost cell service on Monday and it didn't come back until the following Sunday. We've got some 10-mile two-ways and we're going to distribute them to friends and family as an alternate means of communication.

Lesson 10—KEEP CASH ON HAND. When we finally made it into town there was no power and no phones in the businesses—hence no credit or debit cards. My wife and I keep cash on hand just for this reason. We have the advantage of living in a community where everyone knows everyone, so we could in most cases write a check, but in a more heavily populated area that's probably not the case.

Lesson 11—Things can get REAL bad REAL fast. I would say I got a brief glimpse into a total SHTF scenario. It was short, but it was bad. No power, no heat, no fuel, dwindling food supplies, people with no preparation—and it can get scary quick. This was just three or four days. You take these things away for two, three weeks or longer—then what?

Lesson 12—Rural is better (in my opinion) and people are generally good folks. There was a group of us that went around on Tuesday af-

> ternoon and Wednesday on four-wheelers checking on folks. We had to take chainsaws and log chains to clear some stuff, but we did it. There were groups of people that got together with four-wheel drive vehicles to take the elderly to the shelters.
>
> I learned some great lessons, and got a great test run. We're making modifications to the plans based on the lessons. I hate that it happened, but it was an excellent teaching moment.

There is a lot we can learn from this story about how fragile our infrastructure is and how people behave in a crisis. Fortunately for KyFarmer, his family lived in a rural area, so they could handle sanitation and, for the most part, avoid violence. But that's only because he was already so well prepared.

Unless you live in an equally rural area and are equally well prepared, a common storm could put you directly in harm's way.

START PREPPING NOW!

My aim in this book has been to inspire you to become more prepared. As I said at the outset, I'm worried about you.

Now that we're nearing the end of the book, I'm no longer worried that you don't know *why* you should prepare. By now you either understand the many events that can suddenly bring devastating changes to your family's way of life—or you don't.

I hope that you do.

I've tried my best to paint a rational landscape of both what could realistically occur as well as what you can do now to be prepared to protect your family. We hear doomsday scenarios too often and I encourage you to not get caught up in what "might" happen.

Who cares? It's not as if any of us hope to emerge from a catastrophe to be the one who was "right."

So let's focus on what is important. And that's achieving basic preparedness as I've outlined in this book.

Of course, I hope that you'll go through each of the ten steps and not only implement all the recommendations I made, but discover more ways of fulfilling needs on your own and implement those as well. That shows you are thinking for yourself. As a result, you will achieve a level of preparedness that very few people will reach.

While that is my hope, I realize there's a group of you—perhaps even a majority—who feel overwhelmed. You've read all this and you're just not sure where to start. You don't know what to do.

Don't worry about doing everything in this book—in most of the disasters of the past 30 years victims would have been just fine with a bucket of food, a few cases of water and some basic supplies. Soap, batteries, solar charger and some baby wipes.

And don't get trapped by preparedness paralysis—do what you can, *and do it today*.

It all boils down to this—you're either prepared to live, or you're not.

QUICK START GUIDE: GET PREPARED FOR TWO WEEKS

I know—you're impatient. You have the money but not the time. Maybe you just want to be told exactly what to do.

Whatever the reason, if you don't want to go through the 10 Steps and make your own choices, below is my recommendation for what you should do to *quickly achieve emergency preparedness for two weeks*. Since this book listed the 10 steps on the Path to Preparedness in order of importance, this guide follows that sequence.

I am making an assumption that you at least have a working flashlight at home with extra batteries as well as a few butane lighters. If you don't—please, get them. You don't need this book, or me, to tell you that.

Now here's your quick-start guide. It's not the cheapest way to go, but it's a fast way to achieve a basic level of preparedness.

1. Put aside 15 gallons of water per person in your family. That's enough for two weeks. You can purchase gallon jugs or buy a collapsible Coleman five-gallon water container and fill with tap water. Store it in a cool, safe place and don't touch it.
2. Get a tent that you can set up within your house. That way, at least you can all huddle together in a small space to keep warm.
3. Buy one Mountain House Just in Case Essentials bucket for each person in your family, and one Mountain House Just in Case Breakfast Bucket for each person in your family. That will give you emergency food for two weeks. You'll save money by following my recom-

mendations in chapter five, but this way is fast if you just want to check it off the list. In an emergency, don't eat any of this food until you consume other food that may be spoiling, such as that in your refrigerator.

4. Buy an inverter for your car and extension cords so that you can power critical electronic devices, such as lamps and communications equipment. Prioritize your freezer over your refrigerator as described earlier. Keep your vehicle tank full!
5. Buy two boxes of contractor garbage bags, a bag of peat moss and three packs of baby wipes.
6. Put aside (hide) at least $200 in small bills and change, preferably more.
7. Copy all your personal documents and put them in safe storage.
8. Assemble a survival bag as described earlier, *and* a stocked car emergency kit. If you have the money and just want to check this off the list, buy an assembled survival kit such as the one at tinyurl.com/assembledkit.
9. Buy either a Kelly Kettle kit or a Solo Stove and Pot combo. Either way you can create a fire with twigs and sticks lying around, keep warm and use the included pot to heat your freeze-dried meals. No need to store extra fuel.
10. Store 10 extra gallons of gas for your vehicle in five-gallon containers. During an extended outage you'll appreciate having the electricity from your inverter.

It was difficult for me to stop at 10, but let's look back at the list. You now have food, shelter, water, heat, medical and sanitation supplies, extra fuel and an evacuation bag. You even have a flashlight and lighters.

What you may not have is personal protection, such as a firearm. But I had to stop the list at 10—otherwise it would never stop. So be aware that even if you have the items on the list, you may be at risk if there is a widespread emergency.

While following this quick start list doesn't earn you status among the prepping elite, it does put you ahead of at least 98 percent of Americans. And you'll be well prepared for two weeks.

So—congratulations! You're in the top two percent.

BREAK FREE OF THE MATRIX

In closing this book I'd like to share a scene from another film I've enjoyed many times. Released in 1999, The Matrix paints a dystopian view of the future, where humans are slaves to the very technology they created to serve themselves.

The "Matrix" itself is a simulated reality—one in which virtually everyone falsely perceives their lives as productive and worry-free.

Sound familiar?

Very few people have ever awoken to the horrifying truth about the Matrix; that comatose humans are physically plugged into the computer network so their body heat can be harvested to fuel the Matrix itself. In other words, the humans only perceive what the Matrix tells them to perceive.

A gripping scene occurs when Cypher, a character freed from the Matrix, finds himself unable to cope with the harshness of the real world. He agrees to sabotage and murder fellow liberated crewmembers in exchange for being plugged back into the Matrix.

In that scene Cypher sits at a restaurant table across from a malevolent computer agent called Agent Smith. The camera zooms into Cypher's weary face as he cuts into a sizzling steak, raises a juicy bite in front of his mouth, pauses, and says to Agent Smith:

> "I know this steak doesn't exist. I know that when I put it in my mouth, the Matrix is telling my brain that it is juicy and delicious. After nine years, you know what I realize? Ignorance is bliss."

Cypher then savors the bite—the bite that does not exist—and rolls his eyes in ecstasy.

The pace of the movie affords little time to reflect on this scene. If it did, we would conclude that the scene is rather depressing, as it portrays a man who would rather live in a state of blissful coma than be independently free to experience life.

Cypher doesn't want to be in control of his own life and no longer wants to know the truth. He just wants to know that he'll be plugged back into the Matrix, where he is promised a perceived life of importance, without a care in the virtual world. That everything will be okay and the powers that be will take care of him.

Don't be Cypher. Don't be afraid to embrace the real world, where real things happen. Some things are good; most are insignificant. Some are catastrophic events that we don't want to think about.

But the real world is also full of wonderful life—your life, and the lives of those you love.

Don't obsess over fear. Enjoy the present, but prepare for whatever the future holds. And remember:

Being prepared may be the most rational and optimistic gift you can offer your family.

If This Book was Helpful to You

PLEASE leave a review on Amazon!

APPENDIX I

PREPPER CHALLENGES

I've mentioned throughout this book that, when it comes to preparedness, having skills is just as important as having stuff. Below are several challenges that you can practice to become a more prepared person—in good times or in bad.

Most of these challenges cost no money; some cost just a little. They're designed to get you thinking, practicing, and prepared for whatever comes your way.

You don't have to do them all at once, of course, but I do hope that you do them all.

PREPPER CHALLENGE—WATER PURIFICATION

- Purify drinking water with bleach.
- Purify drinking water with calcium hypochlorite.

PREPPER CHALLENGE—SAFE SPOTS

- Find the safe spots in your home for each type of disaster and practice with all family members.
- For example, during an earthquake you would want to practice "drop, cover, and hold on" under a sturdy desk or table.
- During a tornado, you would want to seek shelter in

a tornado shelter, or a lower-level room without windows.

PREPPER CHALLENGE—HOUSE TENTS

- On a very cold winter day, set up a dome or pup tent within your house.

PREPPER CHALLENGE—TEMPORARY SHELTER

- Establish a temporary shelter outside.
- Use simple materials, such as a tarp between trees, or create a simple debris shelter.
- Invite your family or friends to help. Then, have a picnic there to celebrate its completion!

PREPPER CHALLENGE—BUILD A FIRE

- Gather natural materials such as twigs, bark, dry grass, etc.
- Build a fire outside using a natural fire starting method or a magnesium fire starter.
- Add increasingly larger sticks and logs until a safe, roaring fire is controlled.
- Have all family members participate and ensure they know how to do the same.
- Finally, once you're proficient with this, do the same challenge—only with wet wood!

PREPPER CHALLENGE—CAN IT!

- Try canning something.
- If you don't know how, ask someone to help, read a canning manual or watch a reputable YouTube video.
- If you don't have a canner, purchase one if you can or find someone who does have one.

PREPPER CHALLENGE—MENU PLANNING

- Make a disaster menu for your family.
- Be sure it covers two weeks.
- Put the food needed on a shelf and label it "disaster menu." Eat this for two weeks and see how you did with your menu planning.
- Replace items and improve the menu as needed.

PREPPER CHALLENGE - OFF THE GRID LAUNDRY:

- Try doing laundry off the grid!
- In the simplest form, doing laundry off the grid requires a water source, plus, you'll need a family-size washboard, a galvanized bucket, a wringer washer, and of course a clothes line.
- Do a week's worth of laundry without using electricity.
- The exercise will enable you to understand your family's needs in the event you're forced to live off the grid.

PREPPER CHALLENGE—GRID-DOWN WEEKENDS

- Have your family simulate two grid-down weekends.
- Do one in the summer, one in the winter.
- Go from Friday at 6:00 pm to Sunday at 6:00 pm.
- Use painter's tape to tape all light switches off.
- Shut off your house water supply or water breaker, and remember to shut off the water heater.
- Unplug landlines and turn off all cell phones (since cell phones won't work grid down).
- You may use battery-operated devices (such as laptops, tablets, etc.) but must recharge them with alternative energy sources.
- Use only your existing preps for food, fire, etc.

PREPPER CHALLENGE—MEDICAL PREPAREDNESS

- Assemble a first aid/medical kit for home and car.
- Simulate various emergencies with all family members and ensure everyone knows what to do.
- Practice first aid for splints, burns, blisters, open wound care, choking, heart attacks and more.

PREPPER CHALLENGE—THREAT AWARENESS

- Spend a day monitoring your colors throughout the day.
- Write down when you advanced from one color to

the next, and why.

- Pay attention to others you see and make a mental note of what color state they are in.
- Imagine you're a criminal—are the other people heads down and oblivious? Could you take advantage of that?
- Apply what you learn to your own mental alertness routine.

PREPPER CHALLENGE—FIRE EXTINGUISHERS

- Teach each family member how to use the fire extinguisher, and show them where it's kept.
- Fire extinguishers will not work properly if they are not properly charged. Use the gauge or test button to check proper pressure.
- If necessary, you can get training from your local fire department.

PREPPER CHALLENGE—HOME INSPECTION

- Do a home intrusion inspection—where can people get in?
- Where are the bad-guy hiding spots?
- You and family try to gain access to your home to find areas you can improve.

PREPPER CHALLENGE—ESCAPE!

- Determine the best escape routes from your home.
- Try to find two ways out of each room. Try going out windows if it is possible to do safely.

PREPPER CHALLENGE—HAZARD HUNT

- Conduct a home hazard hunt with your family.
- During a disaster, anything that can move, fall, break, or cause a fire is a home hazard.
- For example, during an earthquake a hot water heater or a bookshelf could fall and injure someone.
- Look for electrical, chemical, and fire hazards.
- Contact your local fire department to identify specific home fire hazards.

PREPPER CHALLENGE—SHUT DOWN

- Show each family member how and when to turn off the water, gas, and electricity at the main switches.

PREPPER CHALLENGE—PAYPHONE

- While out and about, make a family game of locating pay phones.
- That way if your family faces a disaster, you all know where they are and why you may need them!

APPENDIX II

Courtesy of FSIS:

REFRIGERATOR FOODS When to Save and When to Throw It Out	
FOOD	**Held above 40 °F for over 2 hours**
MEAT, POULTRY, SEAFOOD Raw or leftover cooked meat, poultry, fish, or seafood; soy meat substitutes	Discard
Thawing meat or poultry	Discard
Meat, tuna, shrimp, chicken, or egg salad	Discard
Gravy, stuffing, broth	Discard
Lunch Meats, hot dogs, bacon, sausage, dried beef	Discard
Pizza, with any topping	Discard
Canned hams labeled "Keep Refrigerated"	Discard
Canned meats and fish, opened	Discard
CHEESE Soft Cheeses: blue/bleu, Roquefort, Brie, Camembert, cottage, cream, Edam, Monterey Jack, ricotta, mozzarella, Muenster, Neufchatel, queso blanco, queso fresco	Discard
Hard Cheeses: Cheddar, Colby, Swiss, Parmesan, provolone, Romano	Safe
Processed Cheeses	Safe
Shredded Cheeses	Discard
Low-fat Cheeses	Discard
Grated Parmesan, Romano, or combination (in can or jar)	Safe

REFRIGERATOR FOODS When to Save and When to Throw It Out	
DAIRY Milk, cream, sour cream, buttermilk, evaporated milk, yogurt, eggnog, soy milk	Discard
Butter, margarine	Safe
Baby formula, opened	Discard
EGGS Fresh eggs, hard-cooked in shell, egg dishes, egg products	Discard
Custards and puddings	Discard
CASSEROLES, SOUPS, STEWS	Discard
FRUITS Fresh fruits, cut	Discard
Fruit juices, opened	Safe
Canned fruits, opened	Safe
Fresh fruits, coconut, raisins, dried fruits, candied fruits, dates	Safe
SAUCES, SPREADS, JAMS Opened mayonnaise, tartar sauce, horseradish	Discard if above 50 °F for over 8 hrs.
Peanut butter	Safe
Jelly, relish, taco sauce, mustard, catsup, olives, pickles	Safe
Worcestershire, soy, barbecue, Hoisin sauces	Safe
Fish sauces (oyster sauce)	Discard
Opened vinegar-based dressings	Safe
Opened creamy-based dressings	Discard
Spaghetti sauce, opened jar	Discard
BREAD, CAKES, COOKIES, PASTA, GRAINS Bread, rolls, cakes, muffins, quick breads, tortillas	Safe
Refrigerator biscuits, rolls, cookie dough	Discard

REFRIGERATOR FOODS When to Save and When to Throw It Out	
Cooked pasta, rice, potatoes	Discard
Pasta salads with mayonnaise or vinaigrette	Discard
Fresh pasta	Discard
Cheesecake	Discard
Breakfast foods—waffles, pancakes, bagels	Safe
PIES, PASTRY Pastries, cream filled	Discard
Pies—custard, cheese filled, or chiffon; quiche	Discard
Pies, fruit	Safe
VEGETABLES Fresh mushrooms, herbs, spices	Safe
Greens, pre-cut, pre-washed, packaged	Discard
Vegetables, raw	Safe
Vegetables, cooked; tofu	Discard
Vegetable juice, opened	Discard
Baked potatoes	Discard
Commercial garlic in oil	Discard
Potato Salad	Discard

FREEZER FOODS When to Save and When To Throw It Out		
FOOD	**Still contains ice crystals and feels as cold as if refriger-ated**	**Thawed. Held above 40 °F for over 2 hours**
MEAT, POULTRY, SEA-FOOD Beef, veal, lamb, pork, and ground meats	Refreeze	Discard

FREEZER FOODS When to Save and When To Throw It Out		
Poultry and ground poultry	Refreeze	Discard
Variety meats (liver, kidney, heart, chitterlings)	Refreeze	Discard
Casseroles, stews, soups	Refreeze	Discard
Fish, shellfish, breaded seafood products	Refreeze. However, there will be some texture and flavor loss.	Discard
DAIRY Milk	Refreeze. May lose some texture.	Discard
Eggs (out of shell) and egg products	Refreeze	Discard
Ice cream, frozen yogurt	Discard	Discard
Cheese (soft and semi-soft)	Refreeze. May lose some texture.	Discard
Hard cheeses	Refreeze	Refreeze
Shredded cheeses	Refreeze	Discard
Casseroles containing milk, cream, eggs, soft cheeses	Refreeze	Discard
Cheesecake	Refreeze	Discard
FRUITS Juices	Refreeze	Refreeze. Discard if mold, yeasty smell, or sliminess develops.

FREEZER FOODS When to Save and When To Throw It Out		
Home or commercially packaged	Refreeze. Will change texture and flavor.	Refreeze. Discard if mold, yeasty smell, or sliminess develops.
VEGETABLES Juices	Refreeze	Discard after held above 40 °F for 6 hours.
Home or commercially packaged or blanched	Refreeze. May suffer texture and flavor loss.	Discard after held above 40 °F for 6 hours.
BREADS, PASTRIES Breads, rolls, muffins, cakes (without custard fillings)	Refreeze	Refreeze
Cakes, pies, pastries with custard or cheese filling	Refreeze	Discard
Pie crusts, commercial and home-made bread dough	Refreeze. Some quality loss may occur.	Refreeze. Quality loss is considerable.
OTHER Casseroles—pasta, rice based	Refreeze	Discard
Flour, cornmeal, nuts	Refreeze	Refreeze
Breakfast items—waffles, pancakes, bagels	Refreeze	Refreeze
Frozen meal, entree, specialty items (pizza, sausage and biscuit, meat pie, convenience foods)	Refreeze	Discard

APPENDIX III

Secondary Preparedness Books

Survival/Preparedness

- First Aid—Responding To Emergencies, by American Red Cross
- How to Survive the End of the World as We Know It, by James Wesley Rawles
- Survival Psychology, by John Leach
- The Boy Scout Handbook, by Boy Scouts of America
- Survival Guide for Beginners, by Vitaly Pedchenko
- Harvesting H2o: A prepper's guide to the collection, treatment, and storage of drinking water while living off the grid, by Nicholas Hyde
- Survival Mom: How to Prepare Your Family for Everyday Disasters and Worst-Case Scenarios, by Lisa Bedford
- The Prepper's Blueprint: The Step-by-Step Guide to Help You Through Any Disaster, by Tess Pennington

Food Storage & Cookbooks

- Ball Complete Book of Home Preserving, by Judi Kingry
- Poverty Prepping: How to Stock up For Tomorrow When You Can't Afford To Eat Today, by Susan Gregersen

- Stocking Up: The Third Edition of America's Classic Preserving Guide, by Carol Hupping
- Putting Food By: Fifth Edition, by Janet Greene
- Root Cellaring: Natural Cold Storage of Fruits & Vegetables, by Mike Bubel

Homesteading Skills

- **How to Make Money Homesteading: So You Can Enjoy a Secure, Self-Sufficient Life**, by Tim Young
- The Backyard Homestead: Produce all the food you need on just a quarter acre!, by Carleen Madigan
- Storey's Basic Country Skills: A Practical Guide to Self-Reliance, by John Storey
- Back to Basics: A Complete Guide to Traditional Skills, Third Edition, by Abigail R. Gehring

Gardening/Farming

- Gardening When It Counts: Growing Food in Hard Times, by Steve Solomon
- Mini Farming: Self-Sufficiency on 1/4 Acre, by Brett Markham
- The New Organic Grower, by Eliot Coleman
- Rodale's Ultimate Encyclopedia of Organic Gardening: The Indispensable Green Resource for Every Gardener, by Fern Marshall Bradley

Medical/First aid

- The Survival Medicine Handbook: A Guide for When Help is Not on the Way, by Joesph Alton
- The Complete Medicinal Herbal: A Practical Guide to the Healing Properties of Herbs, with More Than 250 Remedies for Common Ailments, by Penelope Ody
- Where There Is No Dentist, by Murray Dickson
- Wilderness Medicine, 5th Edition, by Paul S. Auerbach

Off-Grid Living

- When Technology Fails: A Manual for Self-Reliance, Sustainability, and Surviving the Long Emergency, by Matthew Stein
- When The Grid Goes Down: Disaster Preparations and Survival Gear For Making Your Home Self-Reliant, by Tony Nester
- The Prepper's Guide To Grid Down Survival: How To Prepare, Survive And Become Self Reliant If The Lights Go Out & The Water, Gas Or Energy Grid Collapses, by Jim Jackson
- The Prepper's Guide To Off The Grid Survival: An Introduction To Living A Stress Free, Self-Sustaining Lifestyle In Financial Peace, by Brandon Davis

Personal & Home Protection

- Retreat Security and Small Unit Tactics, by David Kobler
- Tappan on Survival, by Mel Tappan
- Survival Guns, by Mel Tappan
- The Secure Home, by Joel Skousen
- When All Hell Breaks Loose: Stuff You Need To Survive When Disaster Strikes, by Cody Lundin
- Principles of Personal Defense, by Jeff Cooper

Foraging, Hunting & Fishing

- Trapper's Bible: Traps, Snares & Pathguards, by Dale Martin
- Complete Guide to Hunting: Basic Techniques for Gun & Bow Hunters, by Gary Lewis
- A Field Guide to Edible Wild Plants, by Lee Allen Peterson

Communication

- Ham Radio For Dummies, by H. Ward Silver
- Personal Emergency Communications: Staying in Touch Post-Disaster: Technology, Gear and Planning, by Andrew Baze

Prepper & Survivalist Fiction

- One Second After, by William R. Forstchen
- One Year After, by William R. Forstchen
- 77 Days in September, by Ray Gorham
- The Road, Cormac McCarthy
- CyberStorm, by Matthew Mather
- Atlas Shrugged, by Ayn Rand
- Brave New World, by Aldous Huxley

Prepper/Self-Sufficiency Books for Children

- Amazing Maisy Hatches an Egg, by Tim Young
- Amazing Maisy and the Honeybee, by Tim Young
- Playful Preparedness!, by Tim Young
- Prepper Pete Prepares, by Kermit Jones Jr.

Other Survival Skills

- The Foxfire Book series, by Eliot Wigginton
- The Home Distiller's Workbook, by Jeff King
- The Outdoor Knots Book, by Clyde Soles
- Bushcraft 101: A Field Guide to the Art of Wilderness Survival, by Dave Canterbury
- Crisis Preparedness Handbook: A Comprehensive Guide to Home Storage and Physical Survival, by Jack A. Spigarelli

- Preparedness Now!: An Emergency Survival Guide, by Aton Edwards
- Making the Best of Basics: Family Preparedness Handbook, by James Talmage Stevens
- Emergency: This Book Will Save Your Life, by Neil Strauss
- How to Stay Alive in the Woods, by Bradford Angier
- Living Well on Practically Nothing, by Ed Romney
- Outdoor Survival Skills, by Larry Dean Olsen
- Camping & Wilderness Survival, by Paul Tawrell
- Just in Case, by Kathy Harrison
- How to Survive Anything, Anywhere: A Handbook of Survival Skills for Every Scenario and Environment by Chris McNab

OTHER BOOKS BY TIM YOUNG

How to Make Money Homesteading (nonfiction)

Digital, print and Audible versions!

Buy How to Make Money Homesteading for your Kindle!

THE ACCIDENTAL FARMERS (memoir)

Digital, print and Audible versions!

Buy The Accidental Farmers for your Kindle!

POISONED SOIL (Fiction)

Digital, print and Audible versions!

Buy Poisoned Soil for your Kindle!

The Amazing Maisy! Picture Books

Digital and print versions!

Find out more at maisybooks.com

ABOUT THE AUTHOR

While flying high over corporate America, Tim Young received a call he couldn't ignore. He shredded his business cards, said goodbye to the conveniences of urban life and become a farmsteader. Along the way Tim became an award-winning cheesemaker and Amazon bestselling author. Today he lives with his wife—the most beautiful and caring woman in the world, his delightful daughter, and a Silky Terrier named Alfie who speaks to him in condescending broken English.

CPSIA information can be obtained
at www.ICGtesting.com
Printed in the USA
FFOW02n1425040116
20105FF